Archaeological Salvage in the
Walter F. George Basin of
the Chattahoochee River in Alabama

David L. DeJarnette

ARCHAEOLOGICAL SALVAGE IN THE WALTER F. GEORGE BASIN OF THE CHATTAHOOCHEE RIVER IN ALABAMA

CONTRIBUTIONS BY

Wesley R. Hurt
Edward B. Kurjack
Fred Lamar Pearson, Jr.

DRAWINGS BY

Oscar W. Brock, Jr.

The University of Alabama Press
Tuscaloosa, Alabama

Reprinted in 2010 by
The University of Alabama Press
Tuscaloosa, Alabama 35487-0380

Manufactured in the United States of America

This study was a cooperative enterprise conducted under an agreement numbered 14–10–0131–799 with the United States Department of the Interior — National Park Service

∞
The paper on which this book is printed meets the minimum requirements of American National Standard for Information Sciences-Permanence of Paper for Printed Library Materials, ANSI Z39.48-1984.

Library of Congress Cataloging-in-Publication Data

DeJarnette, David L. (David Lloyd), 1907–
Archaeological salvage in the Walter F. George Basin of the Chattahoochee River in Alabama / David DeJarnette.
p. cm.
Originally published: University : University of Alabama Press, 1975.
Includes bibliographical references and index.
ISBN 978-0-8173-5644-6 (paper : alk. paper) — ISBN 978-0-8173-8521-7 (electronic) 1. Chattahoochee River Valley—Antiquities. 2. Alabama—Antiquities. 3. Walter F. George Reservoir Region (Ala. and Ga.)—Antiquities. 4. Indians of North America—Chattahoochee River Valley—Antiquities. 5. Indians of North America—Alabama—Antiquities. 6. Excavations (Archaeology)—Chattahoochee River Valley. 7. Excavations (Archaeology)—Alabama. 8. Archaeological surveying—Chattahoochee River Valley. 9. Chattahoochee River Valley—History, Local. 10. Alabama—History, Local. I. Title.
F332.C4D45 2010
975.8—dc22

2010005685

Acknowledgments

The authors wish gratefully to acknowledge the following agencies and institutions without whose assistance and cooperation neither the preliminary survey of the Walter F. George Basin area nor the subsequent excavations would have been possible.

The Research Committee of The University of Alabama in Tuscaloosa financed the preliminary survey that located 124 village sites and 20 mounds and mound groups. The excavations later conducted at many of these sites would have been impossible had it not been for this initial assistance and invaluable insight shown by the committee in financing this survey work.

The United States Department of the Interior-National Park Service provided the funds for the subsequent excavations in the George Basin. This work was a product of the Inter-Agency Archaeological Salvage Program that has been the cooperative effort of all agencies and institutions in the salvage of archaeological, historical, and paleontological resources in advance of civil works construction. John Griffin, then Regional Archaeologist, National Park Service, provided valuable assistance and cooperation throughout the work.

The reports comprising this publication are as originally written in 1949 and 1963. No attempt has been made to revise or update their substance.

Contents

Figures

Tables

Archaeological Salvage in the Walter F. George Basin of the Chattahoochee River in Alabama

Introduction

During the period from June 10 to September 1, 1947, the first phase of the archaeological salvage program, a preliminary archaeological survey of the Chattahoochee Valley covering the area from the Florida border to Phenix City, Alabama, was undertaken by the Alabama Museum of Natural History. The project was under the general direction of David L. DeJarnette, Curator of the Museum and Steve B. Wimberly, Archaeologist, Mound State Monument; field work was conducted by Wesley R. Hurt, Research Associate of the Museum. The survey was financed by the Research Committee of The University of Alabama. Grateful acknowledgment is made to the committee for its assistance.

The area surveyed included, from north to south, Russell, Barbour, Henry, and Houston counties. Although limited time and heavy woods hindered exploration in certain areas of the valley, investigators attempted to find as many aboriginal remains as possible. Since the purpose of the survey was to locate sites that were in danger of submersion by proposed dams on the Chattahoochee River, only a few sites that lay outside the valley were investigated. Rarely did the width of the area to be flooded extend more than 4 miles west of the river.

Original plans were to excavate small, stratigraphic trenches at the more important sites. In the time allotted, however, it was not possible to conduct more than minor excavations at Abercrombie Village (1 Ru 61) near Phenix City, site 1 Ho 6 in Houston County, site 1 He 1 in Henry County, and site 1 Br 14 in Barbour County.

One hundred and twenty-four village sites whose surfaces yielded sufficient sherds for analysis were located by the 1947 survey. In addition, 20 mounds and mound groups and a large flint quarry were found within this area.

Additional related research by Dr. Hurt in the Chattahoochee area included studies of several private collections, although the private collections proved to be of little aid in reconstructing the local cultural history because precise information regarding their recovery was lacking.

The second phase of the archaeological salvage program, the excavation of selected sites, was conducted during 1960, 1961, and 1962. In the summer of 1960 two student field training units, one from The University of Alabama, the other from Florida State University, were employed. During the 1961 and 1962 seasons, a local labor crew was hired and supervised by University of Alabama graduate assistants Edward Kurjack and Oscar W. Brock, Jr., assisted by John Cottier, a student at Auburn University. A list follows of the sites excavated, dates of excavation, and supervisors responsible for the excavation:

Name of Site	*Site No.*	*Date*	*Supervisor*
Shorter Site (Mound)	1 Br 15x1	July 1960	D. L. DeJarnette
Shorter Site (Village)	1 Br 15	June 1960	D. L. DeJarnette
Jackson Site	1 Br 35	August 1960	D. L. DeJarnette
Fontaine Reeves	1 Br 37	Sept., Oct. 1961	Oscar W. Brock, Jr.
McLendon Site	1 Ru 28	August 1960	Lewis Larson
		July, August 1962	Edward B. Kurjack
Abercrombie Site	1 Ru 61	June 1962	Edward B. Kurjack
Patterson Site	1 Ru 66	July 1962	Edward B. Kurjack
Cottonton Site	1 Ru 69	July 1960	Lewis Larson
Spanish Fort Site	1 Ru 101	June 1960	Lewis Larson

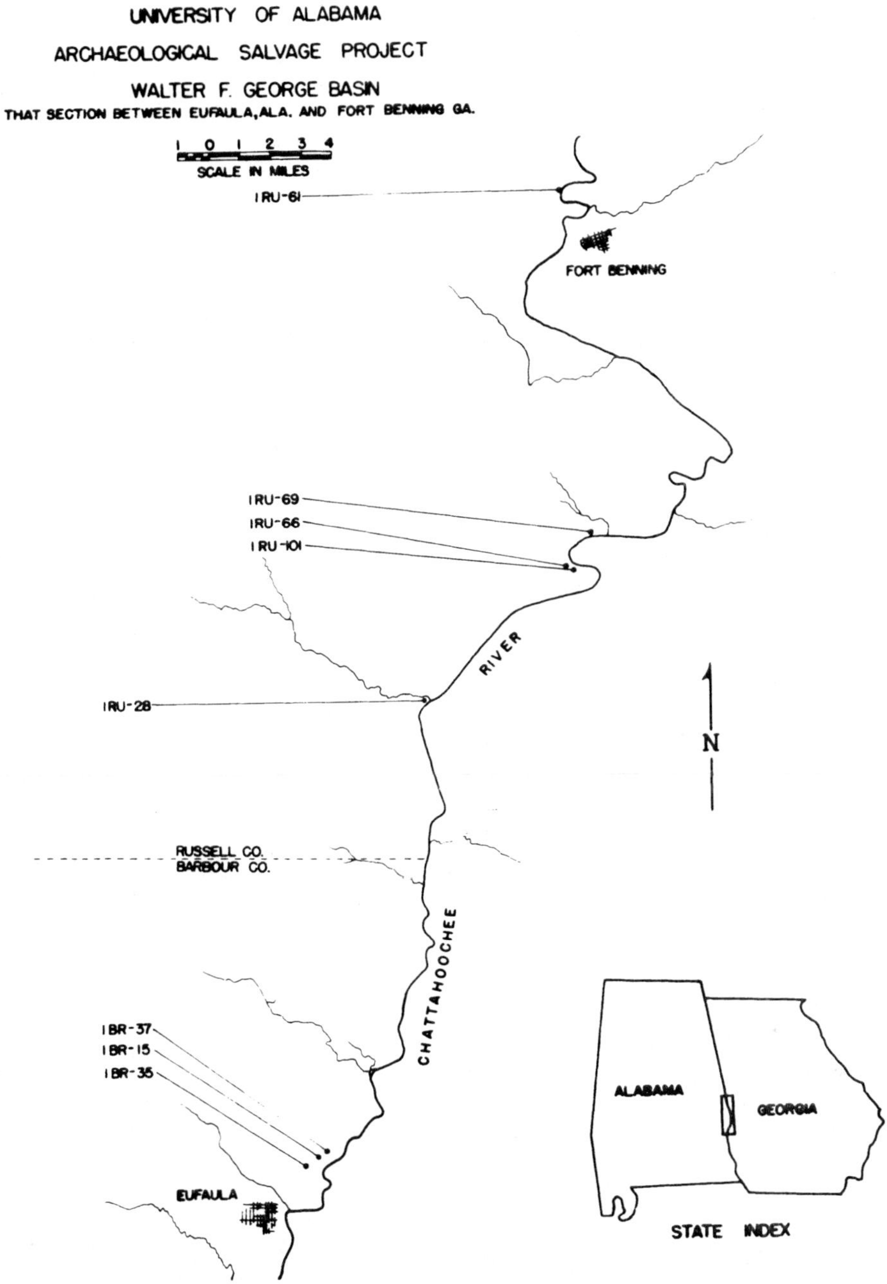

Figure 1 Archaeological Salvage Project – Walter F. George Basin

Part I

THE PRELIMINARY ARCHAEOLOGICAL SURVEY OF THE CHATTAHOOCHEE VALLEY AREA IN ALABAMA

Wesley R. Hurt

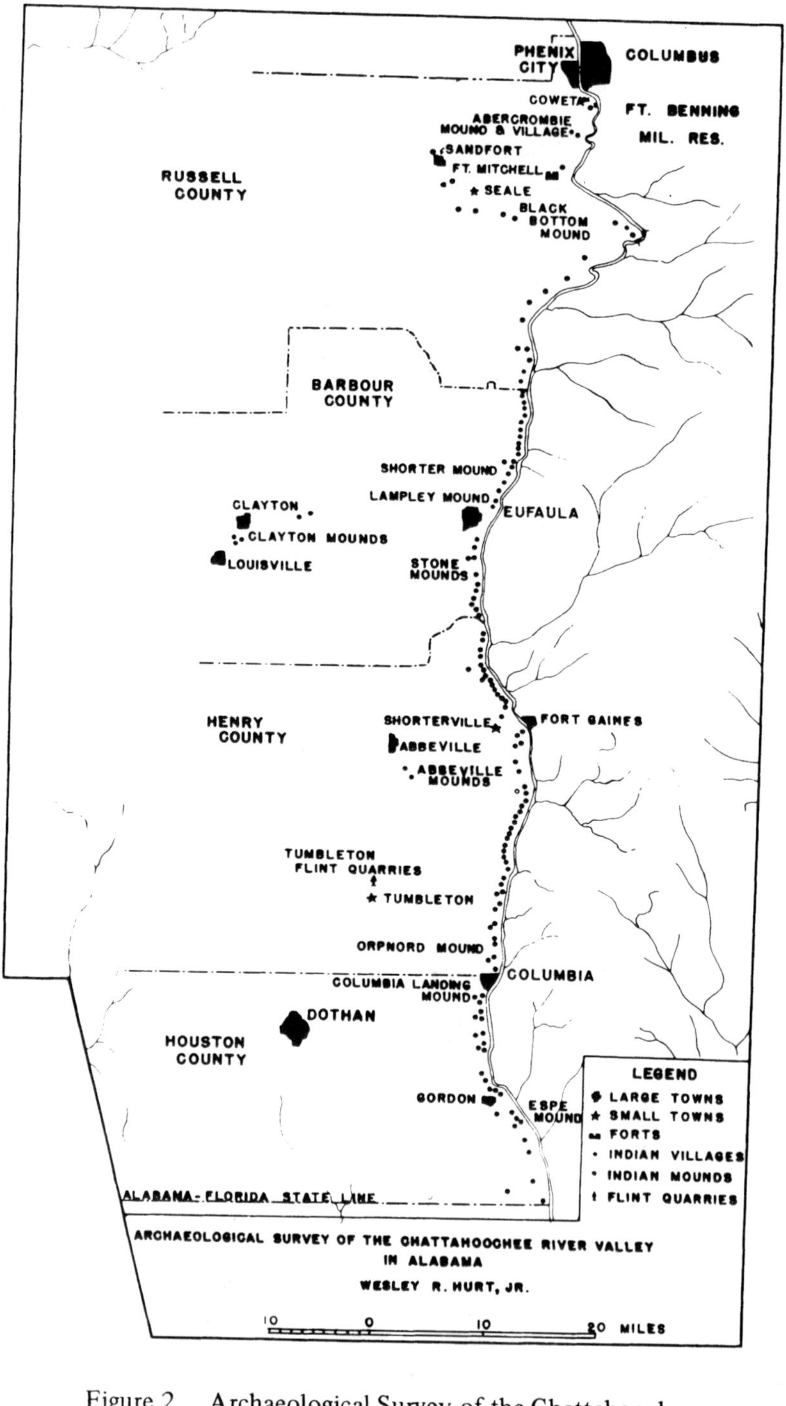

Figure 2 Archaeological Survey of the Chattahoochee River Valley in Alabama

CHAPTER 1

The Historic Tribes in in the Survey Area

INTRODUCTION

In Part I of this publication only those historic tribes of the Chattahoochee that occupied the area covered by the preliminary survey will be considered. The purpose of Part I is to aid in the identification of those sites found in the survey. This attempt to associate historic tribes with archaeological sites involved great difficulties. The early maps depict areas in a most generalized fashion; the names and locations of streams often have changed through time; some areas have been buried by floods or eroded, while others, former town sites, are now covered with dense vegetation. Further complication is caused by the successive, consecutive occupation of a single site by tribes of diverse cultural affiliation. Thus the search for a particular historic site at a particular area usually leads to the discovery of several sites, all of which may fit the vague descriptions left by early explorers.

DESCRIPTION BY EARLY EXPLORERS

Several accounts of the area of the Chattahoochee by early explorers and travelers are available today, but only the journals of Benjamin Hawkins and William Bartram provide enough detail to identify the Indian village sites. In 1785 Benjamin Hawkins was appointed by the Congress of the United States to a commission to negotiate a treaty with the Creeks (Caughey 1938: 27), and in 1796 he became the United States Agent to the Creeks and general superintendent of the Indians south of the Ohio. He resided with the Creeks until after the War of 1812 (Caughey: 55, 95). In 1799 he visited the Chattahoochee Indian villages and left a detailed description of their locations that was published in the *Georgia Historical Society Collections* (1848).

William Bartram was a botanist and naturalist who traveled in the southeastern part of the present United States on an extended trip beginning in 1773. In July he visited several of the Creek towns in the Chattahoochee Valley, but because he visited so few of them, his descriptions are not quite so useful as those of Hawkins (Van Doren 1928: 312-317).

HISTORIC MAPS

Swanton has published in his *Early History of the Creek Indians* (1922) several maps and census lists that depict the historic towns of the Chattahoochee Valley. In this group are:

1. "The Territory between the Chattahoochee and Mississippi River." This map was compiled by Baron de Crenay, Commandant of the Post of Mobile, in March, 1733, and reproduced in the French archives for Peter J. Hamilton, who published it in the second edition of his "Colonial Mobile," 190-196.
2. "The Southeastern Part of the Present United States," from the Popple Map of 1733.
3. "The Southeastern Part of the Present United States," from the Mitchell Map of 1755.
4. "Part of Purcell Map" compiled not later than 1770 in the interest of British Indian Trade by John Stuart, H. M. Superintendent of Indian Affairs. The original is in the Edward E. Ayer Collection, Newberry Library, Chicago, Illinois.
5. "Towns of the Creek Confederacy," from the Early Map of Georgia, 1818.

In addition Swanton has reproduced a map titled, "The Distribution of Indian Tribes in the Southeast," about the year 1715, redrawn from a blueprint of the original among the British Archives. Swanton states that the disposition of the tribes indicates that the map was prepared shortly before the outbreak of the Yamasee War of 1715. While this map is of little value in locating any of the village sites in Alabama, it does indicate the location of many of the tribes before they moved to the Chattahoochee after the Yamasee War. Swanton also published, in this same volume, two maps (Plates 1 and 2) that are a compilation of his work on the location of the Indian tribes of the Southeastern United States (1922).

TOWN LISTS AND CENSUS

In his volume on the *Early History of the Creek Indians,* Swanton includes a chart giving the comparative town censuses of the Southeastern Indians (1922, 434-437). Data are included from the Spanish census of 1738; French census of 1750; Census of 1760; Census of 1761; Taitt, 1772; Maybury, 1792; Hawkins, 1799; and U. S. Census of 1832.

A COMBINED LIST OF THE POST-1700 VILLAGES

In Swanton's (Charts IA-ID) the various Indian village sites visited by Bartram and Hawkins and the sites on the maps described above are summarized. Numbers indicate successive positions of the tribes. Letters are used where tribes separate into two or more bands. Taken together, these lists indicate the following village sites along the Chattahoochee:

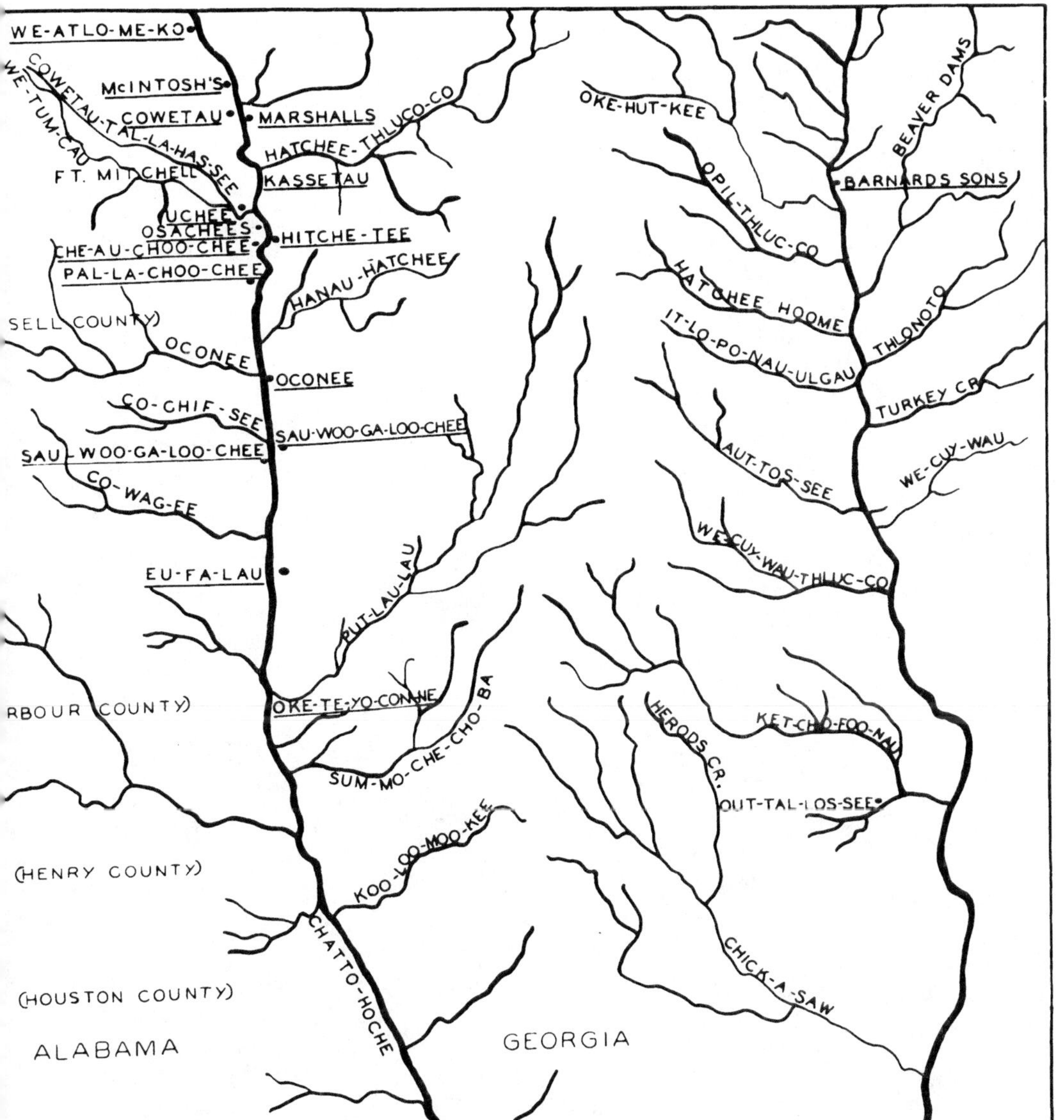

Adapted from "Towns of the Creek Confederacy" as shown on the Early Map of Georgia, 1818 (Swanton, 1922).

Figure 3 Location of Historic Creek Villages

Alabama side, moving down river from Phenix City	*Georgia side, moving down river from Columbus*
Kawita (3)	Kasihta (7)
Coweta Tallahassee (Likatcka 2)	Kasihta (6), later Chickasaw (3b)
Kawita (2)	
Yuchi (3b)	Likatcka (1)
Osochi (4)	Okmulgee (2)
Chiaha (5a)	Hitchiti (4)
Chiahutci	
Westo (Yuchi 4b)	
Apalachicola (4)	Oconee (2a)
Kolomi (2)	Sawoklutci
Atasi (2)	Tamahita (2)
Tuskegee (2a)	Eufaula Hopai
Kasihta (5)	Okitiyagani
Okmulgee (1)	Eto-Husse-Wakkes
Sawokli (3)	
Chiaha (4a)	
Okawaigi	
Hogologee (Yuchi 3b)	
Tamali (3)	
Shawnee (3a)	
Hitchiti (3)	
Yamasee (4b, 6b)	

LOCATION OF TOWNS VISITED BY HAWKINS IN 1799

Only the journals of Benjamin Hawkins and William Bartram describe the towns of the Chattahoochee in detail sufficient to make a reasonable identification of the various post-1700 archaeological sites. While this report is not primarily concerned with the identification of sites in Georgia, identification of a few of them is necessary for the location of the Alabama sites.

The following identification of three sites visited by Hawkins in 1799 begins with sites in the area of Phenix City, Alabama, and continues with succeeding downriver sites. The first Creek villages to be identified are, in Swanton's terminology, Coweta (3) and Coweta Tallahassee (or Likatcka: 2). The following is an abstracted version of Hawkins' description of these three towns (Hawkins 1848: 52-56).

> *Cow-e-tuh,* on the right bank of Chat-to-ho-che, three miles below the falls, on a flat extending back one mile . . . the fields of the settlers who reside in the town, are on a point of land formed by a bend of the river . . .
>
> *Cow-e-tuh Tal-lau-has-see* . . . It is two and a half miles below Cowetuh, on the right bank of the river. In going down the path between the two

TABLE 1

HISTORIC CREEK TOWNS IN AREA 1947 SURVEY OF CHATTAHOOCHEE TAKEN FROM EARLY MAPS AND DOCUMENTS, ALABAMA SIDE, NORTH TO SOUTH

1733 De Crenay Map		1733 Popple Map		1755 Mitchell Map		1770 Purcell Map	
Original Spelling	Probable Identification	Original Spelling	Probable Identification	Original Spelling	Probable Identification	Original Spelling	Probable Identification
		Coetaus	Coweta				
						Cussita	Kasihta
		Westos	Yuchi	Euchees	Yuchi	Woristo	Yuchi
		Culloowes		Westos	Yuchi		
		Achina	Chiaha	Hogoleega's	Yuchi		
		Hogologes	Yuchi				
				Okmulgee	Okmulgee		
Chouakale	Sawokli					Swagala	Sawokli
Tamatle	Tamali						
		Savanoes	Shawnees				
						Hicheta	Hichiti
						Taussie	Tawasa

TABLE 2

ALABAMA TRIBES–HISTORIC CREEK TOWNS IN AREA 1947 SURVEY OF CHATTAHOOCHEE TAKEN FROM EARLY MAPS AND DOCUMENTS, ALABAMA SIDE, NORTH TO SOUTH

Towns visited by William Bartram, 1778		Towns visited by Benjamin Hawkins, 1799		Villages Noted on Map of Georgia, 1818		Swanton Map Compiled in 1919
Original Spelling	Probable Identification	Original Spelling	Probable Identification	Original Spelling	Probable Identification	
		Cow-e-tuh	Coweta	Cowetau	Coweta	Kawita (3)
		Cow-e-ta	Likatcka	Cowetau		Likatcka (2)
		Tal-lau-has-see		Tallahas-see	Likatcka (2)	Kawita (2)
Uche	Yuchi	Uchee	Yuchi	Uchee	Yuchi	Yuchi (3b)
				Osachees	Osachi	Osochi (4)
						Chiaha (5a)
		Oose-oo-che	Osochi	Che-au-choo-chee	Chiahutci	Chiahutci
Chehaw	Chiaha	Che-au-hau	Chiaha			
		Che-au-hoo-chee	Settlement of Chiaha			Westo (Yuchi 4b)
				Pal-la-choo-choo	Apalachicola	Apalachicola (4)
						Kolomi (2)
						Atasi (2)
						Tuskegee (2a)
						Kasihta (5)
						Okmulgee (1)
				Sau-woo-ge-loo-chee	Sawokli	Sawokli (3)
						Chiaha (4a)
						Okawaigi
						Hogologee (Yuchi 3b)
						Tamali (3)
						Shawnee (3a)
						Hitchiti (3)
						Yamasee (4b, 6b)

TABLE 3

HISTORIC CREEK TOWNS ON THE GEORGIA SIDE–CHATTAHOOCHEE FROM NORTH TO SOUTH

1733 De Crenay Map		1733 Popple Map		1755 Mitchell Map		1770 Purcell Map	
Original Spelling	Tribal Affiliation	Original Spelling	Probable Identification	Original Spelling	Probable Identification	Original Spelling	Probable Identification
		Gitafes	Kasihta	Cowetas	Coweta		
		Osochi	Osochi				
		Oakmulgo	Ocmulgee			Ocmulgia	Okmulgee
		Echetes	Hitchiti				
		Ocones	Oconee			Oconee	Oconee
Tamaitaux	Tamahita						
Chya	Chiaha	Atafes	Astas	Attassees	Atasi	Swagalat-chie	Sawoklutchi
						Lower Eufalla	Eufaula
		Jalkages	Tuskegee	Jackeeges	Tuskegee		
				Cullomas	Kolomi		
				Chiha	Chiaha		
				Savannas	Shawnee		
						Palachoc-ota	Apalachicola

TABLE 4

GEORGIA TOWNS TAKEN FROM MAPS AND DOCUMENTS—CHATTAHOOCHEE FROM NORTH TO SOUTH

Towns visited by Benjamin Hawkins, 1799		Villages noted on Maps of Georgia, 1818		Swanton Map Compiled in 1919
Original Spelling	Probable Identification	Original Spelling	Probable Identification	
Cus-se-tuh	Kasihta	Kussetau	Kasihta	Kasihta (7)
				Kasihta (6) later
Chickasaw	Chickasaw			Chickasaw (3b)
				Likatcka (1)
Ocmulgee	Okmulgee			Okmulgee (2)
Hit-che-tee	Hitchiti	Hitche-tee	Hitchiti	Hitchiti (4)
O-co-nee	Oconee	Oconee	Oconee	Oconee (2a)
Sau-woo-ge-lo	Sawoklutci	Sau-woo-ga-loo-chee	Sawoklatci	Sawoklutci
				Tamahita (2)
		Eu-fa-lau	Eufaula	Eufaula Hopai (Sawokli Village)
				Okitiyagani
		Oke-te-yo-co-ne	Okitiyagani	Eto-Husse-Wakkes

> towns, in half a mile cross Kotes-ke-le-jau, ten feet wide, running to the left is a fine little creek sufficiently large for a mill, in all but the dry seasons. On the right bank, enter the flat lands between the towns. These are good, with oak, hard-shelled hickory and pine: they extend two miles to Che-luc-in-ti-ge-tuh, a small creek five feet wide, bordering on the town. The town is half a mile from the river, on the right bank of the creek; it is on a high flat, bordered on the east by the flats of the river, and west by high broken hills; they have but a few settlers in the town . . .
>
> They estimate their number of gun men at one hundred; but the agent has ascertained, by actual enumeration, that they have but sixty-six, including all who reside here, and in the villages belonging to the town.

In the adjacent area on the Georgia side of the Chattahoochee, Hawkins described the location of Kasihta (6), which later became Chickasaw town (3b).

> *Cus-se-tuh*; this town is two and a half miles below Cow-e-tuh Tal-lau-has-see, on the left bank of the river . . . In descending the river path from the falls in three miles you cross a creek running to the right, twenty feet wide; this creek joins the river a quarter of a mile above the Cowetuh town house. . . .
>
> The creek has its source nearly twenty miles from the river, and runs nearly parallel with it till within one mile of its junction; there it makes a short bend round north, thence west to the river; at the second bend, about two hundred yards from the river, a fine little spring creek joins on its right bank. . . .
>
> The flat of good land on the river continues two and a half miles below this creek, through the Cussetuh fields to Hat-che-thluc-co. At the entrance of the fields on the right, there is an oblong mound of earth; one quarter of a mile lower there is a conic mound forty-five yards in diameter at the base, twenty-five feet high, and flat on the top . . . the river makes a short bend round to the right, opposite this mound, and there is a good ford just below the point . . . the river from this point comes round to the west, then to the east . . .
>
> The land back from the fields to the east, rises twenty feet, and continues flat for one mile to the pine forest; back of the fields adjoining the rise of twenty feet, is a beaver pond of forty acres, capable of being drained at the small expense of labor; the large creek bounds the fields and the flat land to the south.
>
> Continuing on down the river from the creek, the land rises to a high flat, formerly the Cussetuh town, and afterwards a Chickasaw town. This flat is intersected with one branch. From the southern border of this flat, the Cussetuh town is seen below, on a flat, just above flood mark, surrounded with this high flat to the north and east, and the river to the west . . . (Hawkins 1848: 57-58).

The reconstruction of Hawkins' trip in this area is more difficult than would appear at first glance because of changes in topography, streams, and names of streams. The following reconstruction would appear most probable:

The Trip on the Alabama side of the Chattahoochee.

(1) *Falls.* The falls visited by Hawkins were probably those that lie in the general area where Falls Creek enters the Chattahoochee on the west bank south of Phenix City.

(2) *Coweta.* Described as being 3 miles below the falls. In this area are sites 1 Ru 8 and 1 Ru 9. Site 1 Ru 9 is the larger and was probably the main village. Site 1 Ru 8 has the Ceramic Complex H and site 1 Ru 9 has the Complex I.

(3) *Kotes-ke-le-jau Creek.* Described as a half mile below Coweta. This is the exact location from site 1 Ru 9 of the modern creek called Cochgalachee Creek.

(4) *Che-luc-in-ti-ge-tuh Creek.* Described as 2 miles south of Kotes-ke-le-jau Creek. The small creek joining the Chattahoochee at Abercrombie Landing is at this distance.

(5) *Coweta Tal-lau-has-see.* A half mile from the river on the right bank of Che-luc-in-ti-ge-tuh Creek. In this location is a medium-sized site, 1 Ru 11, which has a mixture of Ceramic Complexes F (Fort Walton), G (Lamar), and H (Ocmulgee Fields). The Fort Walton and Lamar component at the site was probably earlier in time. Site 1 Ru 11 (1 Ru 62) is much smaller than 1 Ru 9, which agrees with the 1832 Census of 896 for Coweta and 438 for Coweta Tallahassee (listed as Broken Arrow). At the mouth of this creek at Abercrombie Landing is the large site 1 Ru 10 (1 Ru 61), described by C. B. Moore in his 1905 trip (Moore 1905: 449-450). It is very doubtful that this was ever the site of Coweta Tallahassee. The location does not fit Hawkins' description. At the site is a large mound that should have attracted Hawkins' attention, for he refers to other mounds in his journals. In addition, the pottery complex indicates a predominance of Fort Walton and Lamar and only a slight amount of Complex H (Ocmulgee Fields). [See Chapter 2 for description of pottery complexes.] Without doubt, however, many of the intrusive burials at 1 Ru 10 (1 Ru 61) belong to a late Creek occupation of the surrounding area.

Sites on the Georgia Side of the Chattahoochee

(1) *Falls of Chattahoochee.* In the absence of contradictory data, assumed to be the falls described by Hawkins on his trip down the west side of river. They are located south of Phenix City, Alabama.

(2) Hawkins mentions encountering a creek on the east bank of the Chattahoochee, 3 miles below the falls. He states that it originates nearly 20 miles from the river, runs parallel with it until nearly within its bounds; there it bends north and then west to the river. On the right bank a small stream enters. The only stream south of the area of Columbus, Georgia, fitting this description is now called Bull Creek. Hawkins noted that this creek entered the Chattahoochee a quarter of a mile upriver from the Coweta town house. This is an exact descrip-

tion of the geographical relationship between the mouth of Bull Creek and site 1 Ru 9, which supports the hypothesis that this site was the location of Coweta.

(3) *Mounds.* Two and a half miles below the creek (probably Bull Creek) Hawkins observed an oblong mound and a quarter mile downriver, a conical mound. No mounds remain there, but records indicate their presence in this area in the 1800's. C. B. Moore states that on the property of J. Kyle, of Willet, Georgia, in view of the river, is the remnant of a mound that repeated floods have washed away (Moore 1907: 450). His map places this mound in the general area under discussion. Apparently the Kyle Mounds were washed away by the flood of 1886 and were in the vicinity of the Woolfolks village site. Hawkins states that the mounds were near the place where the Chattahoochee bends to the right, the location of the Woolfolks site. According to Peter Brannon, "About 1885 the river began to wash into what was later known as the Kyle mound and the celebrated flood of 1886 carried away much of it." (1925: 44)

(4) *Large Creek South of Mounds.* Hawkins mentions a creek entering the river below the mounds. The present-day Upatoi Creek matches this description.

(5) *Old Cussetuh and Chickasaw Town.* Hawkins says this village was downriver from the mound creek on a high flat. This description roughly coincides with the area of the Sergeants' Club on the Fort Benning Military Reservation. Mrs. Isabel Patterson reported a site in this area, which was not visited by the writer.

(6) *Cussetuh Town.* Located on a flat that could be seen from the southern border of the high flat occupied by the Chickasaw town. The large village site that was destroyed by the construction of the airport at the Fort Benning Military Reservation fits this description.

PETER A. BRANNON'S RECONSTRUCTION OF HAWKINS' JOURNEY

Peter A. Brannon locates the sites of Coweta and Coweta Tallahassee visited by Hawkins in this manner:

> The town [referring to Coweta Tallahassee] was on the right bank of the Chattahoochee River, one-half mile inland, on a beautiful, level plateau, extending about one and one-half miles back to the hilly wooded country, two miles below the present post office of Brickyard and immediately on the left of the Columbus and Andalusia branch (the old Mobile and Girard division) of the Central of Georgia R.R. It is on the plantation of the late Major Moffit Flournoy, in Sec. 23, Township 16, Range 30 East, Russell County, Alabama. The site is 8½ miles, as the crow flies and about 15 miles by river, below Columbus, and about 2½ miles above old Fort Mitchell, immediately on the right bank of and one-half miles above the mouth of, a small stream now called Broken Arrow Creek (Brannon 1930:37).

The site in the area described by Brannon was given the number 1 Ru 25 by the 1947 survey.

To support his thesis that site 1 Ru 25 was the Coweta Tallahassee visited by Hawkins, Brannon gives this reconstruction: "In conclusion, going down the path between the two points, as Hawkins did in 1797, we still cross, in a half mile, Kooteshalijau Creek, and after passing the mound, arrive at Che-luc-in-te-ge-tuh, now known as Broken Arrow, a name given it by the wife of the former owner of the property (Mr. Flournoy) . . . " (Brannon 1930: 45).

In Brannon's reconstruction of Hawkins' trip the following assumptions are made:

(1) The present mouth of Cochgalachee Creek (Hawkins' Kotes-ke-le-jaù Creek) does not correspond to the mouth of the creek during Hawkins' time. To support this assumption, Brannon states, "Originally the stream entered the Chattahoochee at what we know at the present time as the Fitzgerald mound near Bickerstaff's Brickyard of today [mound at Abercrombie Village, 1 Ru 61]. By the construction of the brick manufacturing establishment in the vicinity of the original Abercrombie Brickyard (now the Bickerstaff Brickyard), the flow of the stream has been turned so that it enters the Chattahoochee higher up." (Brannon 1924: 67)

(2) That Abercrombie Village, 1 Ru 10 (1 Ru 61) occupies the general area of Coweta visited by Hawkins.

(3) That site 1 Ru 25, at the mouth of Broken Arrow Creek, was Coweta Tallahassee visited by Hawkins.

(4) That the creek now called Broken Arrow Creek was the Che-luc-in-ti-ge-tuh visited by Hawkins.

If we accept Brannon's reconstruction of Hawkins' journey, the following difficulties are encountered:

(1) The falls near Falls Creek, by Phenix City, Alabama, are not the falls described by Hawkins. As Hawkins stated that they were 3 miles upstream from Coweta (Abercrombie Village, according to Brannon), they would have been in the vicinity of the Kyle Mounds. Locating the falls at this point makes reconstruction of Hawkins' description of his trip down the *east* side of the Chattahoochee impossible. Hawkins' description of the location of the streams and their physical features on the east bank of the Chattahoochee simply makes little sense unless the falls were located in the general area of Phenix City.

(2) If the reconstruction given previously of Hawkins' trip down the east bank is accepted as correct, then site 1 Ru 9 rather than Abercrombie Village 1 Ru 10 (1 Ru 61) was the "Coweta" visited by this explorer. It should be noted that the Coweta town house was said by Hawkins to be a quarter mile below the mouth of a creek that appears to have been Bull Creek. In this area is site 1 Ru 9 and not 1 Ru 25.

Brannon's location of Kasihta on the east side of the river is identical with location of this village derived by the writer's reconstruction of the east river journey of Hawkins. Brannon states in regard to Kasihta:

> The old Cusseta town of later days was located on the hill up the river from Hall's landing and where the present Fort Benning Military Reservation quarters now sit. History records the fact that the town house was about where the General's official residence is now located. The old town was doubtless located about the lower landing and opposite the river ferry crossing at Fort Mitchell (Brannon 1925: 44).

(3) Abercrombie Village, 1 Ru 10 (1 Ru 61) is plainly a Ceramic Complex G (Lamar) and does not have any indications of the European trade we would expect if it were the late Historic site of Coweta, as stated by Brannon (1924-b: 67-68). Without doubt many of the intrusive burials at the site are Historic. If 1 Ru 10 (1 Ru 61) is the site of Coweta, it seems strange that Hawkins did not mention the large mound, for he describes in detail the other mounds in the valley. It should be noted that Hawkins describes Coweta as being a half mile upriver from the mouth of a stream. An intensive search in the area upriver from the stream entering the river at Abercrombie Village did not reveal a trace of a village site. Site 1 Ru 9, however, is a half mile upriver from the mouth of Cochgalachee Creek.

(4) The "Early Map of Georgia, 1818," published by Swanton gives a fairly detailed location of the Lower Creek Towns (Swanton, 1922, Plate 9). On this map the town of Coweta (spelled Cowetau) is placed opposite a stream that can be no other than Bull Creek. This location is the site of present-day 1 Ru 9. Coweta Tallahassee is placed upriver from Fort Mitchell, at the large bend in the river, which must be the bend by Abercrombie Landing. It should also be noted that it is upriver on the opposite bank of the river from Hatcheethluc-co Creek (which can be no other than the present Upatoi Creek). Site 1 Ru 25, identified as that of Coweta, is downriver on the opposite bank between Fort Mitchell and the mouth of Upatoi Creek.

In summary, the writer believes that the sites visited by Hawkins in 1799 on the Alabama side of the Chattahoochee include 1 Ru 9 as the location of Coweta, and 1 Ru 11 (1 Ru 62) the village of Coweta Tallahassee.

Yuchi. In 1799 Hawkins visited a Yuchi town on the Chattahoochee, which he describes as follows:

> U-chee: is on the right bank of Chat-to-ho-che, ten and a half miles below Cow-e-tuh Tal-lau-has-see, on a flat of rich land, with hickory . . . the flat extends from one to two miles back from the river. Above the town, and bordering on it, Uchee creek, eighty-five feet wide, joins the river (Hawkins 1848: 61).

Bartram visited this village in 1778, and says of it: "The Uche town is situated in a vast plain, on the gradual ascent as we rise from a narrow strip of low ground immediately bordering on the river: it is the largest, most compact, and best situated Indian town I ever saw . . ." (Van Doren 1928: 312).

While Hawkins' description of the location of the Yuchi town appears to be clear, a search on both banks of the present-day Uchee Creek was unsuccessful. The area at the time of the 1947 survey was so overgrown with heavy underbrush and trees that only test trenching could have found the site.

Osochi. Hawkins visited this site in 1799 and has left this brief description: "*Oose-oo-che*: is about two miles below Uchee, on the right bank of Chat-to-hoche; they formerly lived on Flint river, and settling here, they built a hot house in 1794; they cultivate with their neighbors, the Che-au-haus, below them, the land in the point" (Hawkins 1848: 63). A search for the village of Osochi was unsuccessful because the area was too heavily overgrown with timber and vegetation. It should be noted that the present-day town of Oswatchee preserves the name of the old Creek village but probably not the original location, for it is set back from the river 4 miles from the bank while Hawkins describes the site to be on the right bank of the Chattahoochee.

Chiaha. The village of Chiaha was visited by Hawkins in 1799. He gives this as the location of Chiaha:

> *Che-au-hau*; called by the traders Che-haws, is just below, and adjoining Oose-oo-che, on a flat of good land. Below the town, the river winds round east, then west, making a neck or point of one thousand acres of cane-break, very fertile, but low, and subject to be overflowed; the land back of this, is level for nearly three miles, with red, post, and white oak, hickory, then pine forest (Hawkins 1848: 63).

This description is consistent with the location of site 1 Ru 12, which has been partially destroyed by a gravel pit at the Bonacre Landing Strip. The ceramics of this site are puzzling in that of a total of 895 sherds found at the site, 894 were a rough variety of Lake Jackson Plain and one was Ocmulgee Fields Incised. No other site found in the 1947 survey had such a high percentage of plainwares. The presence of the Ocmulgee Fields Incised sherd indicates a post-Lamar date for site 1 Ru 12, a date that supports its identification as the old location of Chiaha.

Chiahutchi. Chiahutchi was a small settlement of Chiaha mentioned by Hawkins in his notes of 1799. He refers to it in these terms:

> *Che-au-hoo-che*; (little Che-au-hau;) is one mile and a half west from Hit-che-tee, in the pine forest, near Au-he-gee; a fine little creek, called at its junction with the river, Hit-che-tee; they begin to fence and have lately built a square. (Hawkins 1848: 64).

Apalachicola. One of the most important of the Lower Creek towns was Apalachicola. Bartram visited the village in 1777 and provided the following account:

> After a little refreshment at this beautiful town (Yuchi), we repacked and set off again for the Apalachucla town, where we arrived after riding over a level plain, consisting of ancient Indian plantations, a beautiful landscape diversified with groves and lawns. . . .
>
> One day the chief trader of Apalachucla obliged me with his company on a walk of about a mile and an half down the river, to view the ruins and site of the ancient Apalachucla: it had been situated on a peninsula formed by a doubling of the river . . . (Van Doren 1928: 313).

Hawkins gives this description of Apalachicola as it appeared in 1799: "Pa-la-chooc-le; is on the right bank of Chat-to-ho-che, one and a half miles below Che-au-hau, on a poor, pine barren flat; the land back from it is poor, broken, pine land . . ." (Hawkins 1848: 65).

Site 1 Ru 18, east of the Holy Trinity Church, is located almost exactly where the town of Apalachicola was during Hawkins' visit. In addition, site 1 Ru 18 is approximately 1.5 miles from the peninsula, south of the town visited by Bartram. There is thus little reason to question this identification. The sherds are of Ceramic Complex H, and the presence of Ocmulgee Fields Incised with the absence of Lamar Complicated Stamped types indicated a post-Lamar date, appropriate if this site is the Apalachicola visited by Hawkins and Bartram.

Sawokli. Sawokli was another Lower Creek town visited by Hawkins in 1799. He has left this account:

> Sau-woo-ge-lo is six miles below O-co-nee, on the right bank of the river (the Chattahoochee), a new settlement in the open pine forest. Below this, for four and a half miles, the land is flat on the river and much of it in the bend is good for corn. Here We-lau-ne (yellow water), a fine flowing creek, joins the river; and still lower, Co-wag-gee (partridge), a creek sixty yards wide at its mouth . . . (Hawkins 1848: 65-66).

The location of the Creek town of Oconee on the Georgia side of the river is important in determining the site of the Sawokli. Describing this village, Hawkins wrote, "*O-co-nee* is six miles below Pa-la-chooc-le, on the left bank of Chat-to-ho-che" (Hawkins, 1848: 65). If we add Hawkins' distances, we arrive at these figures: from the mouth of Ihagee Creek to Apalachicola, 1.5 miles; from Apalachicola downriver to Oconee, 6 miles; from Oconee downriver to Sawokli, another 6 miles. These distances place Sawokli 13.5 miles downriver from the mouth of Ihagee Creek.

In the northern part of Barbour County is a creek now called Walaune. However, this creek is at least 20 miles downriver from Ihagee Creek, making it doubtful that this was the We-lau-ne Creek referred to by Hawkins. About 14 miles above the present-day Walaune Creek in Barbour County, Barbour Creek enters the river. On the right bank is a medium-sized site, 1 Br 30, which contains Ceramic Complex I. This complex is similar to that of all the other Historic Creek sites that have been tentatively identified in this report, but differs from some of them in its lack of Ocmulgee Fields Incised. Near the left bank of the present-day Walaune Creek is site 1 Br 21, which also contains Complex I ceramics.

Both Brannon and Battle believe that site 1 Ru 20, east of Cottonton, Alabama, is the site of the Sawokli visited by Hawkins. Brannon states:

> Hatchechubbee . . . Also the name of the stream draining much of the western section of the county . . . Both in the Lower Creek and in the Upper Creek Nations are to be found streams bearing this name . . . According to Hawkins, this village [Hatchechubbee] was settled from Sawokli, a large town located at the mouth of this stream. In Hawkins' report, this stream is referred to as Welaune and is set down as Yellow Water Creek . . . We are positive that the stream now bearing this name, is the one referred to by Hawkins, as Welaune Creek, though at the present time, there is such a stream name in Barbour County, five miles south of the line. (Brannon 1924a:7-8)

H. B. Battle agrees with Brannon's location of Sawokli. He states: "It is one mile S.W. of the point where this railroad (S.A.L.Ry.) crosses the Chattahoochee river, and at the mouth of Hatchechubee Creek" (Battle 1921: 81). Battle adds this information:

> The site of Sawokli was found by an exploring party on July 22, 1921. . . . being guided by 'Uncle' Orv Philips, the aged negro . . . The field perhaps ¾ mile long and ½ mile wide, is now being used as a pasture land, although it is well adapted to corn and cotton. Our location of the town site was positive as we were extremely fortunate in having Orv present to show us the exact spot. His memory was distinct in recalling incidents occurring in the life of the town prior to 1836. (Battle 1921: 82)

Site 1 Ru 20 and nearby site 1 Ru 21 on the north side of the railroad crossing are in the area described by Brannon and Battle. Both of these are large sites and probably represented one village. Ceramic Complex I is found at 1 Ru 21; Complex H is present at 1 Ru 22. The only difference between Complexes H and I is the presence of Ocmulgee Fields Incised in the former. This complex is closely related to that of both 1 Br 21 and 1 Br 30, the other probable sites of Sawokli.

If we accept Brannon's and Battle's identification of 1 Ru 20-21 as the site of the Sawokli visited by Hawkins, we cannot explain the fact that these sites at the mouth of Hatchechubbee Creek are only about 8 miles below the mouth of Ihagee Creek. Hawkins described Sawokli as being 13.5 miles distant. Apparently Brannon and Battle rely on Orv Philips' identification of sites 1 Ru 20-21 as Sawokli. Sawokli was abandoned in 1836. At the time of Battle's and Brannon's visit to the site with Philips in 1921, 85 years had elapsed from the time that the town was last inhabited and 122 years since Hawkins' visit. The writer is inclined, in view of these facts and the great discrepancy in distance, to believe that site 1 Ru 20-21 could not have been the location of Sawokli at the time of Hawkins' visit. It is possible, however, that the location of Sawokli at a later date was moved upriver to the mouth of Hatchechubbee Creek and site 1 Ru 20-21, as identified by Orv Philips, represents this hypothetical later location.

Site 1 Br 30, on Barbour Creek, Barbour County, agrees most closely with Hawkins' location of the town in 1799. Site 1 Br 21 in Barbour County on Walaune Creek and sites 1 Ru 20-21 are less likely possibilities. As their pottery complexes indicate contemporaneity and close relationships, these sites can be expected to yield similar cultural analyses, regardless of which is the site of Hawkins' Sawokli.

Only those Lower Creek villages described by Bartram and Hawkins can be located with any certainty because the early maps and descriptions are too vague; however, it is possible to make certain guesses as to some others.

Westos. On the Popple Map of 1733 is a town called Westos, located south of Cowetans (Coweta) and Culloowes (Kasihta) on the right bank of the Chattahoochee. Many sites would fit this description. The Mitchell Map of 1755 locates Westos south of the Euchees.

Kolomi, Atasi, Tuskegee (5), and *Okmulgee (1)* were located on the Chattahoochee, but Swanton gives no exact information (Swanton 1922: 189). In this area are several sites that have the same ceramic complexes as those visited by Bartram and Hawkins. These include sites 1 Ru 22, 1 Ru 21, 1 Ru 20, 1 Ru 4, 1 Ru 5, and 1 Ru 2, which contain Ceramic Complexes H and I.

Chiaha (4a), Okawaigi, Hogologee, and *Tambi (3),* according to Swanton, were located south of Sawokli on the Alabama side of the Chattahoochee (Swanton 1922: Plate 2 and 142-143). Early maps, such as the Mitchell, De Crenay, Popple, and Purcell, are not helpful in locating these sites. Ceramic Complexes I and H appear on many sites in the area, but it would be impossible to associate these sites with this list of Indian villages.

Shawnee (3a), Hitchiti (3), and *Yamasee (4b,6b)* Swanton lists in the lower area of the Chattahoochee (Swanton 1922: Plate 2). He calls attention to the resemblance of the Creek name Yamasee to the present-day Omusee Creek, which enters the Chattahoochee close to the north border of Houston County. If one judges by their pottery complexes, however, there are no sites at the mouth of this creek that are late enough to be Yamasee Village. Sites 1 Ho 5, 12, and 17 show slight amounts of the Chattahoochee Brushed ware characteristic of Ceramic Complexes I and H found on the Historic sites to the north. Site 1 He 1, a few miles north of the mouth of Omusee Creek in Henry County, also has pottery of this type. All these sites are multi-component, and the majority of their pottery is from prehistoric cultures.

THE REMOVAL OF THE CREEKS TO OKLAHOMA

On March 24, 1832, a treaty made at Washington by General Lewis Cass, Secretary of War, and certain Indian chiefs provided for the removal of the

Creek Indians to a territory west of the Mississippi. On September 17, 1835, the government arranged for several contractors to effect the movement of the Creeks from Alabama to a territory west of Arkansas. Another contract was made by the government with the Alabama Emigrating Company on May 13, 1836, to remove the remaining Indians. By December, 1836, all but 4,106 had been removed, and the remainder were subsequently transported from this state to the Indian Territory between the Canadian and Arkansas rivers.

CHAPTER 2

Study and Analysis of the Survey Material

ARCHAEOLOGICAL COMPLEXES

NON-CERAMIC ARTIFACTS FROM THE CHATTAHOOCHEE AREA

The oldest artifacts found in the general area of the Chattahoochee Valley of Alabama are a series of fluted projectile points from sites lacking pottery. They are similar in form and technique of manufacture to other pre-ceramic projectile points from other areas.

The largest group of projectile points described in this report came from the private collection of Dr. R. B. McCann of Seale, Alabama. Most of the artifacts in his collection were found on the surfaces of sites in fields and on hilltops near Seale in an area about 10 miles west of the Chattahoochee River. The area is wooded and is drained by many small streams. In the same region are several small sites of Ceramic Complex A (a variant of the Deptford focus) and Ceramic Complex B (a variant of the Swift Creek focus). The McCann Collection of about ten thousand points included four fluted points of the type frequently called "Eastern Folsomoid."

These fluted points are described below. The numbers assigned are for reference only.

FA1 (Figure 14) Large point made of cream-colored chert. Has a slightly constricted neck and slightly eared base that displays very little lateral expansion. Finely retouched edges and bases, with grinding at edges. Shallow fluting extends a third of the length along one side. Tip and one side of base broken. Point 61 mm. long, 34 mm. wide, 5 mm. thick.

FA2 (Figure 14) Much shorter than FA1, with constricted neck and base. Base and lower half of sides retouched with ground edges. Single fluted channel on one face, double channel on other. Channeling half length. Point 40 mm. long, 25 mm. greatest width, 8 mm. thick.

FA3 (Figure 14) Similar to those above except base displays more lateral expansion. Made of yellow chert. Fine flaking with retouched edges and ground base. On one face is a central flake, a third the length, framed on each side by a slightly shorter and narrower flake. Point 42 mm. long, 30 mm. wide, and 6 mm. thick.

FA4 (Figure 14) Narrower point than those above with a shorter-eared base. Made of brown chalcedony. Finely chipped with finely retouched edges and a single groove extending a quarter the length on one side. Point 35 mm. long, 20 mm. wide, 4 mm. thick.

A fluted projectile point similar to type FA1 (Figure 14) was found near the race track south of Columbus, Georgia. It differed from the others in that the fluting extended half the length on both sides. The channel was of the single-flake type. Arthur R. Kelly illustrated a projectile point from Macon, Georgia, with an almost identical outline (Kelly 1938: Figure 5). This type of point resembles the Clovis fluted point, which is found associated with mammoth bones in the Southern Plains (Howard 1935: Plate XXIV, Group II). Both the typical Clovis point and the Folsom point generally have a single channel flute, while the Enterline fluted point from the Pennsylvania area, described by John Witthoft (1954: 271-273), has the triple flute characteristic of FA3.

Other projectile points from the McCann Collection that may be pre-ceramic in age are three examples catalogued as IA-A (Figure 14). These points have more pronounced basal ears, which project laterally to a greater extent than those on the fluted points. On one of the points a deep flake has been removed to form the concave base, causing superficial resemblance to a fluted point. The finely retouched edges and bases have a lenticular, or diamond-shaped, cross section. One point displays alternating beveled sides. Range in measurements: Example A, 38 mm. long, 16 mm. wide, 7 mm. thick; Example B, 40 mm. long, 21 mm. wide, 8 mm. thick; Example C, 45 mm. long, 24 mm. wide, 7 mm. thick. These points are made from quartzite or chert. Josselyn (1952: 65-66) illustrated points of this type from non-ceramic sites in the vicinity of Birmingham, Childersburg, Scottsboro, and other areas of Alabama. Robert H. Lister, who has analyzed the distribution of the fish-tailed or eared-based projectile points in the West, concludes that " . . . there does seem to exist a consensus of opinion which would place the artifact in question between the Folsom and recent horizons." (Lister 1953: 265)

Another series of points from the McCann Collection are more triangular in shape with concave bases that frequently are ground. These points are often two-thirds as wide as they are long. Classified Type TC-C (Figure 14), these points differ from the concave-based, triangular point in the ceramic sites because of their larger size, greater width, and basal grinding. Quite a few of them

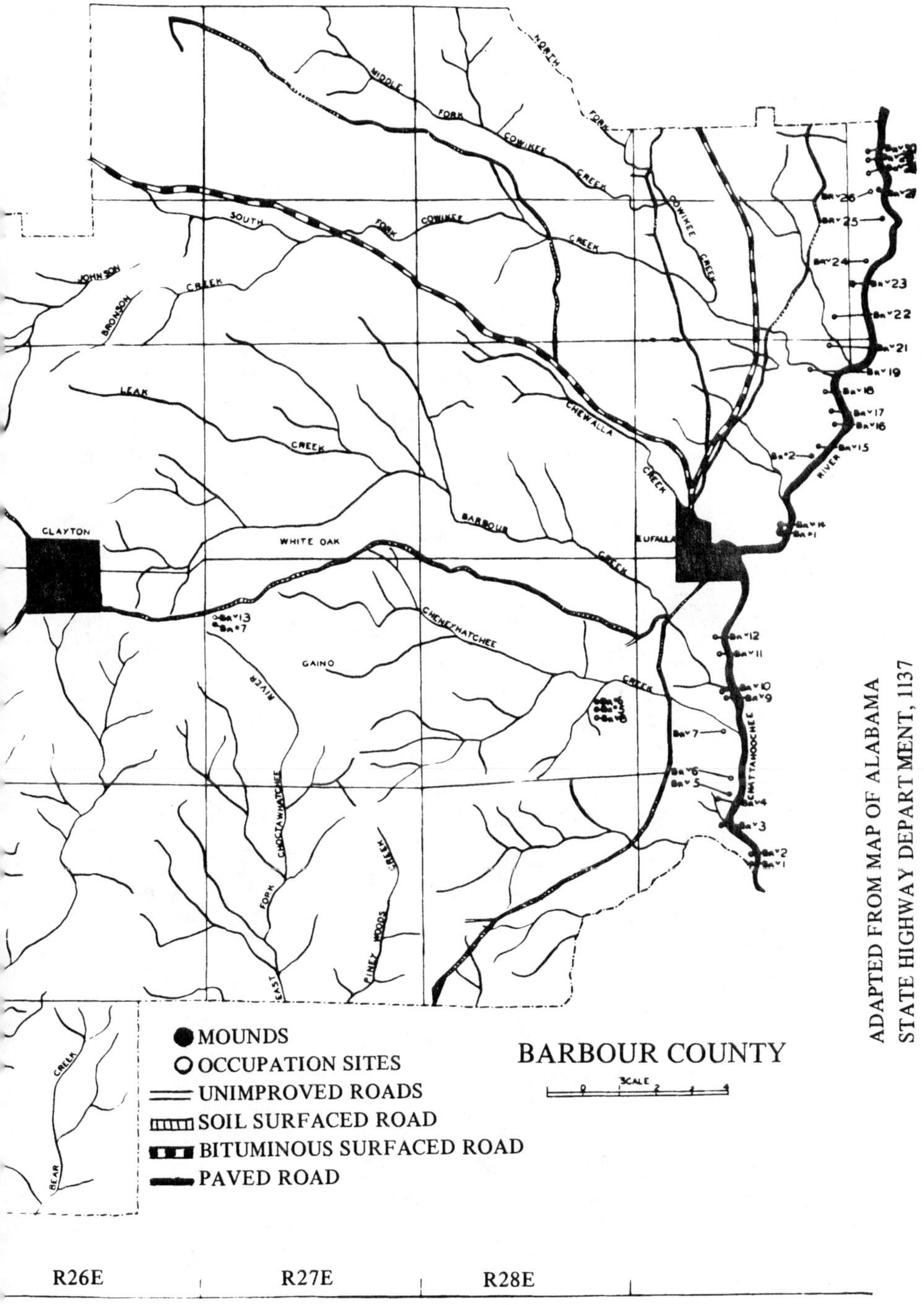

Figure 4 Barbour County Sites Found in 1947 Survey

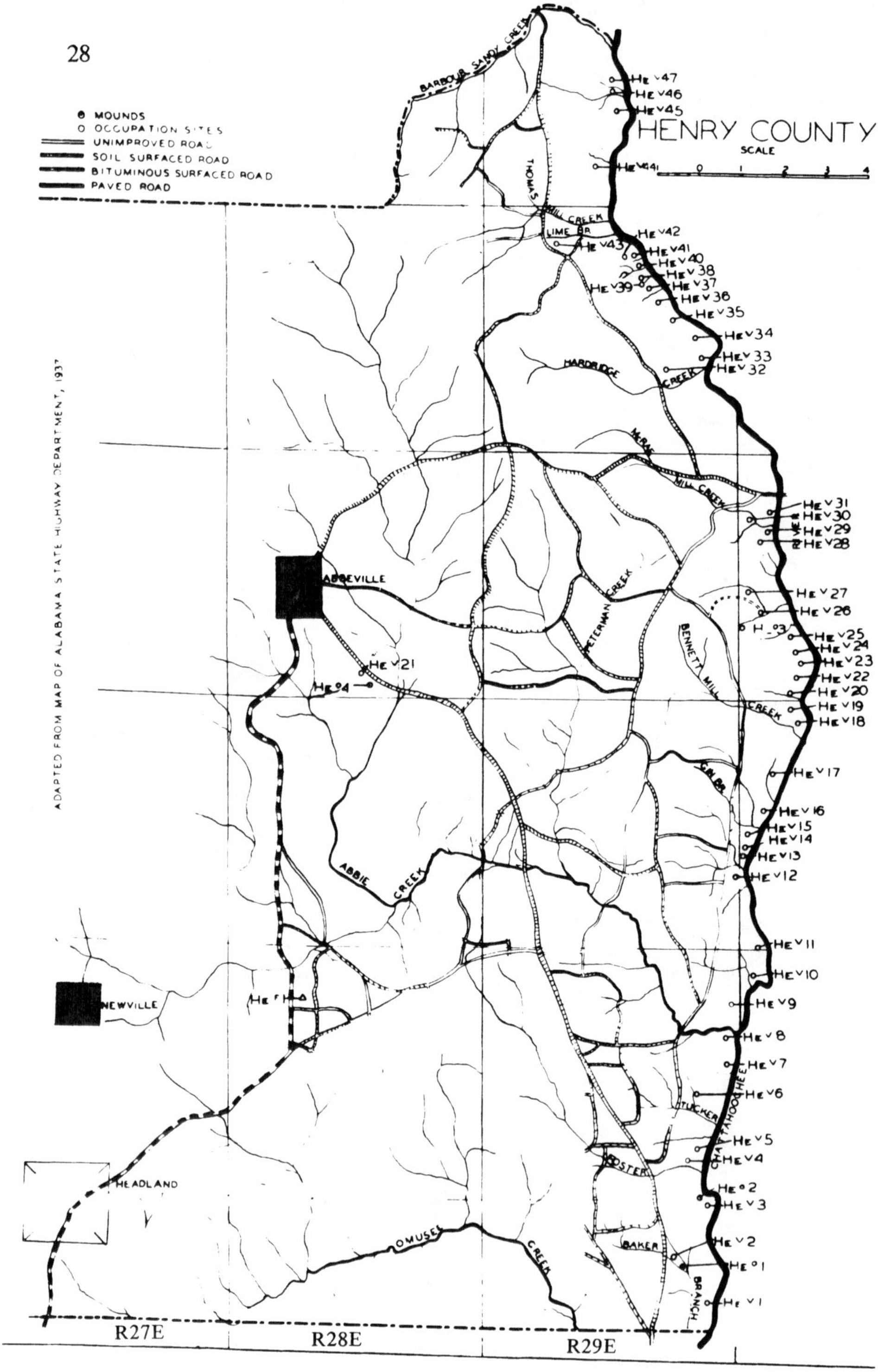

Figure 5 Henry County Sites Found in 1947 Survey

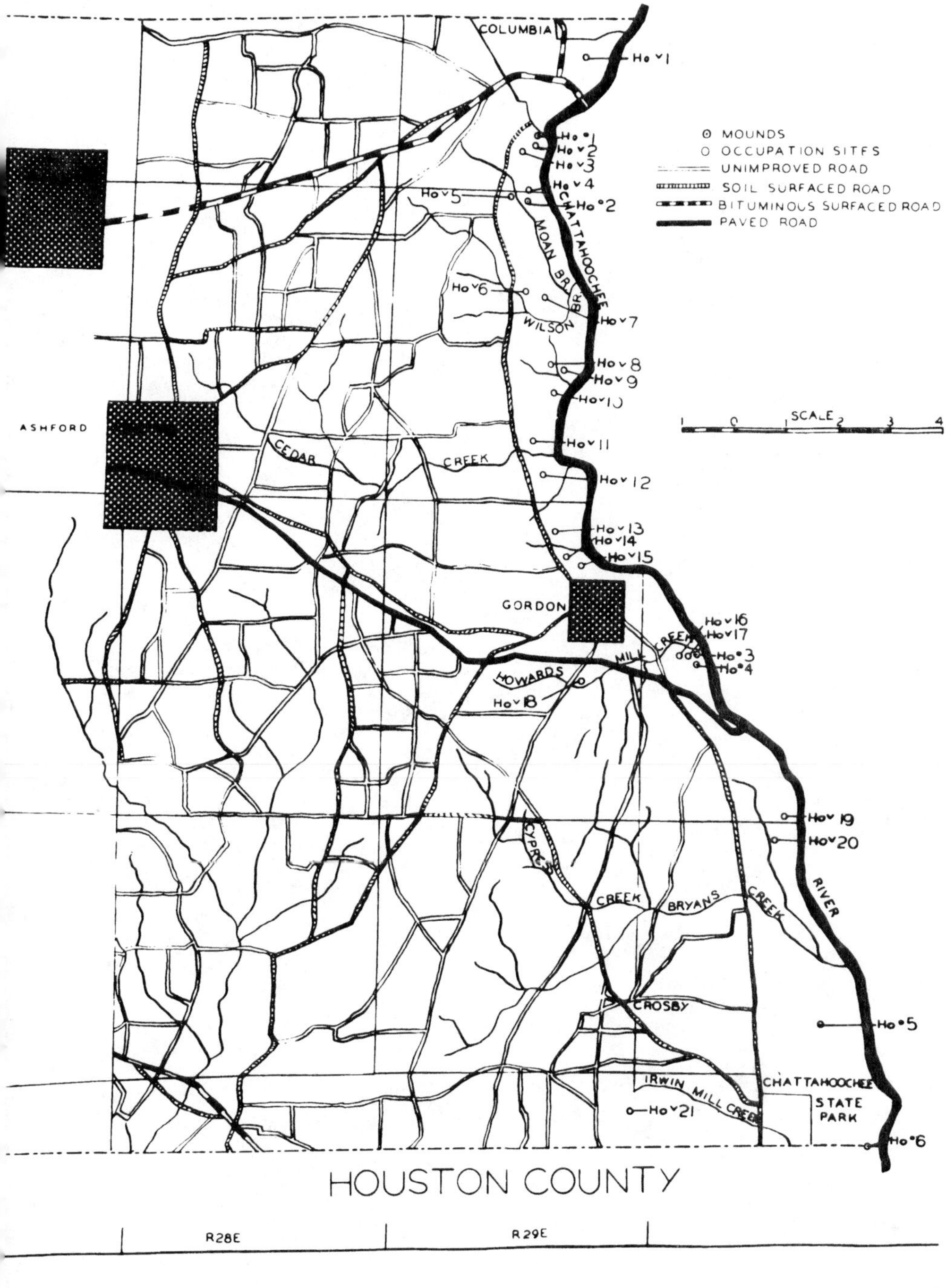

Figure 6 Houston County Sites Found in 1947 Survey

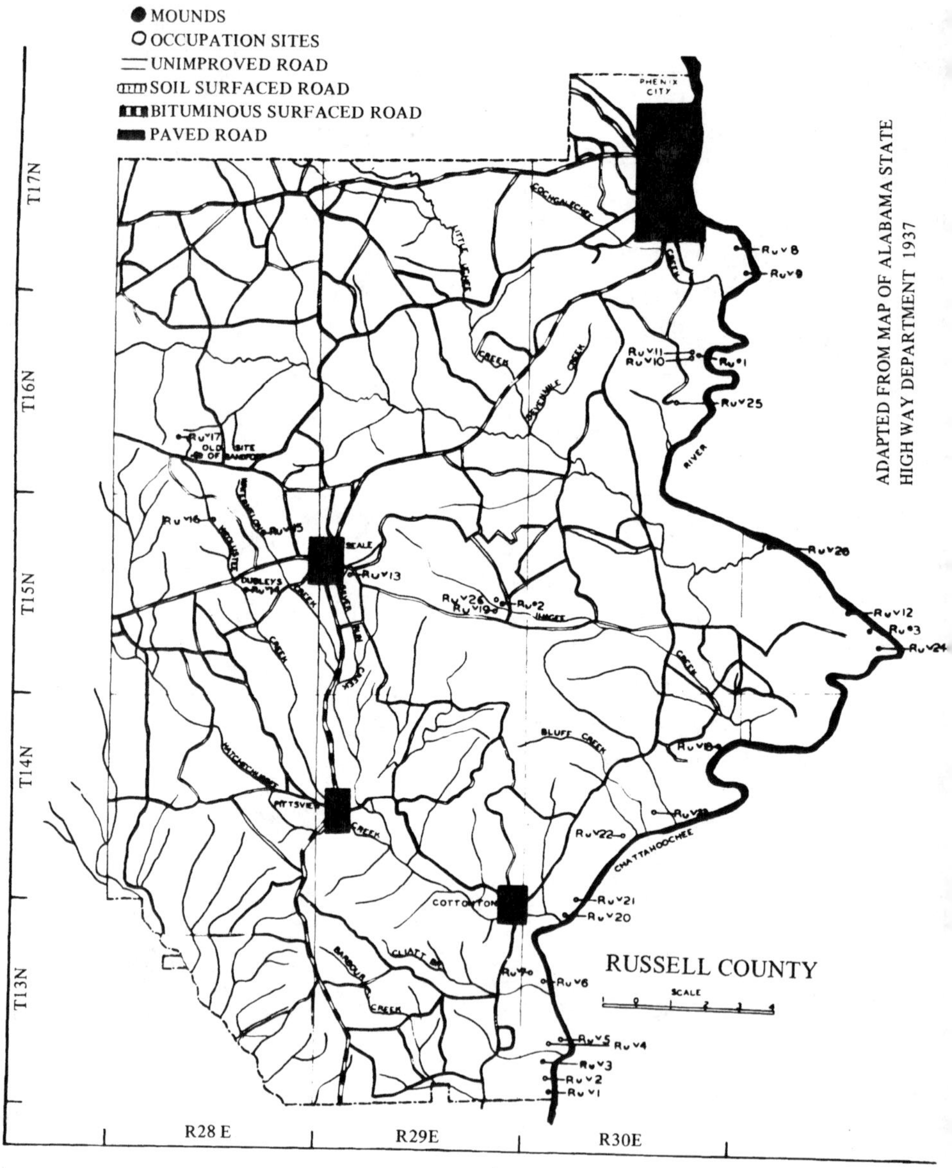

Figure 7 Russell County Sites Found in 1947 Survey

are similar to points illustrated by Josselyn from the Kleine Site on the Tennessee River, Alabama (Josselyn 1952: 69), where many fluted points have been found.

A variant of TC-C is IB. These points have a more pronounced pentagonal shape (Figure 14). The blade is more constricted than the base. The bases and edges are retouched. They are about 40 mm. long, 25 to 30 mm. wide at the base, and 4 mm. thick. A further variant is IS, a triangular point, with slight side notches and a concave base. To form the concave base, a deep flake has been removed, creating a channel that superficially resembles a flute. IS is 35 mm. long, 29 mm. wide, 9 mm. thick.

CERAMIC COMPLEX A

Sites were classified Ceramic Complex A on the basis of sherds collected from the surface. The following pottery types make up the complex: Deptford Linear Check Stamped, Deptford Bold Check Stamped, Dunlap Fabric Marked, and Seale Plain. The presence of the Deptford pottery types indicates that this complex is related to the Deptford focus of the northwest Florida coast, where it is the earliest cultural unit associated with pottery (Willey and Woodbury 1942: 240-241). Along the coastal area of eastern Georgia and South Carolina, the Deptford focus postdates the Savannah River focus, the latter being the earliest ceramic complex (Caldwell 1952: 312-315).

Sites with the Ceramic Complex A are very rare in the Chattahoochee Valley. Only one single-component site, 1 Ho 19, was found by the 1947 survey. This site is in Houston County on a small sandy ridge, west of a tree-lined stream that flows into the Chattahoochee River. This site is very small, and is scattered over a surface of about 50 by 75 yards in an area often caught by floods.

Another single-component site, 1 Ru 16, was investigated. This site, however, was located 4 miles northwest of Seale, Alabama, or about 13 miles west of the Chattahoochee. The site lies in the woods on top of a high ridge along the east side of Weolustee Creek. Sherds and stone chips were found in small concentrated areas that probably were a series of small camps. The overall length of the site was about 150 yards. Sherds were very rare, but stone chips were common.

At site 1 Ru 15, a mile west of 1 Ru 16, Deptford Bold Check Stamped sherds were recovered, although pottery of Ceramic Complex B (Swift Creek) was also present on this site. The site lies in an old field along a gully east of Watermelon Creek. On the south it is bordered by woods and a swamp. The site has been highly eroded.

Table 5 indicates the distribution of pottery types from the single-component Ceramic Complex A sites.

Because of the small size and number of Ceramic Complex A sites, few artifacts were found, although stone chips were plentiful on the sites. Below is a description of the artifacts from the single component sites.

Nutting stone. Made of a fine-grained slab of mica schist, circular in shape. On

TABLE 5

Sherd Count – Ceramic Complex A

	1 Ru 16	1 Ho 19
Deptford Linear Checked Stamped	2	1
Deptford Bold Checked Stamped		3
Dunlap Fabric Marked	4	
Seale Plain	26	8
Other	2	
Total	34	12

one side was a small depression for a nut. Length, 88 mm.; width, 63 mm.; thickness, 28 mm. From site 1 Ru 16.

Projectile point, type SPb. Stemmed point with concave base. One example from 1 Ru 16.

Projectile point, type SF. Constricted base point. One example from 1 Ru 16.

Drill point. Drill in form of a long, thin cone with rounded base. One example from 1 Ru 16.

CERAMIC COMPLEX B

Ceramic Complex B sites were classified on the basis of sherds collected from the surfaces of sites found in the 1947 survey. The following pottery types comprise the complex: Swift Creek Complicated Stamped, Seale Plain, Weeden Island Plain, Carrabelle Incised, West Florida Cord Marked, and Columbia Utility. Lack of intensive excavations in the Complex B sites in the Chattahoochee Valley renders it difficult to place this material precisely in time.

The presence of the Swift Creek Complicated Stamped sherds indicates an affinity to the Santa Rosa-Swift Creek complex of the northwest Gulf Coast of Florida as this complex is defined by Willey and Woodbury (1942: 341-342; Willey 1949: 100). Several of the pottery types assigned by Willey, however, have not been found in the Ceramic Complex B of Alabama. These include St. Andrews Complicated Stamped (Early Variety), New River Complicated Stamped, and Crooked River Complicated Stamped (Early Variety). In addition, Willey assigns to the Santa Rosa-Swift Creek complex of Florida the early vari-

TABLE 6

Sherd Count – Ceramic Complex B (Swift Creek)

Pottery Type	1He38	1Br6	1Br10	1Br19	1Br26	1Ru13	1Ru15	1Ru17	1Ru19
Carrabelle Incised			1				1		
Carrabelle Punctate	1								
Columbia Red-Utility			4				4		23
Comer Punctate									
Keith Incised									
Mound Field Net Marked									
Seale Plain								3	
Stallings Plain					1				
Swift Creek									
Complicated Stamped	1	1	2	4	7	9	2		1
Weeden Island Red Filmed					4				
Weeden Island Plain	32		9	12	106	63	9	6	23
West Florida Cord Marked								1	
Wilson Check Stamped									
Other	3	4	9	2			9		
TOTAL	37	5	25	18	118	72	25	10	47

ety of Swift Creek Complicated Stamped (Willey 1949: 100). Rims and designs of Ceramic Complex B types more closely resemble the middle-to-late types defined by Kelly from the Macon Plateau area of Georgia (Kelly 1938: 28, Plate 11A). The late variety of Swift Creek Complicated Stamped in Florida is assigned by Willey to the Weeden Island I period (Willey 1949: 101).

Though rare in the Chattahoochee and surrounding area, Ceramic Complex B sites are about four times more numerous than Complex F sites, there being eleven Ceramic Complex B sites.

The sites are all small to medium in size and are mainly in the two northernmost counties, Barbour and Russell counties. Only one site, 1 He 38, was in Henry County, and it was toward the north end. Several of these sites occur in the same general area as the Ceramic Complex A site in the hills near Seale, Alabama. Sherds are more common on the Complex B sites than on the Complex A sites, as are stone chips and artifacts.

Artifacts from Ceramic Complex B

Mortar. Made of a triangular slab of ferruginous sandstone that has been pecked into shape. The circular grinding depression on one side is 6 mm. deep. Length, 153 mm.; width, 145 mm.; thickness, 45 mm. From site 1 Ru 15.

Nutting Stone. Made of a slab of ferruginous sandstone. On one side are two circular depressions for holding nuts. The other side is pecked with a circular grinding depression. Length, 130 mm.; width, 123 mm. From 1 Ru 15.

Gorget. Made from a polished piece of sericite schist. Rectanguloid shape with the two narrower ends rounded. Two holes pierce the artifact, each offset from center. Fragmentary. Length, 85 mm.; width, 50 mm.; thickness, 2 mm. From site 1 He 21.

Gorget. A fragment of a gorget that probably had a shape similar to that of the above. Portions that may have borne holes were missing. Made of polished sandstone. Length, 38 mm.; width, 44 mm.; thickness, 6 mm. From site 1 Br 19.

Mortar. Made of a triangular slab of ferruginous sandstone that has a pecked surface. The circular grinding depression on one side is 45 mm. deep. From site 1 Ru 15.

Projectile Points.

NAb1 - leaf-shaped point, two examples. (Figure 13)

NB1 - large, triangular point, one example. (Figure 13)

SAb - side-notched, triangular point, two examples. (Figure 13)

SAd - side-notched, triangular point with round base, seven examples. (Figure 13)

SF - constricted stemmed point, two examples. (Figure 16)

SH - small point with square stem, two examples. (Figure 16)

TC - large triangular point with concave base, one example. (Figure 14)

Drills. DA - cylindrical drill, two examples. (Figure 15)

Knives. M3 - large, asymmetrically shaped blade, two examples. (Figure 15)

M4 - large, ovoid, bifaced blade, one example.

CERAMIC COMPLEX C

Ceramic Complex C sites were classified on the basis of sherds collected on the surface in the 1947 survey. This ceramic complex is more complicated than that of Complex B, and is readily distinguished by the presence of Wilson Check Stamped, in addition to Swift Creek Complicated Stamped types.

The pottery types included in this complex are Seale Plain, Weeden Island Plain, Weeden Island Punctated, Weeden Island Red Filmed, Swift Creek Complicated Stamped, Stallings Plain, Carrabelle Incised, Wilson Check Stamped, Mound Field Net Marked, West Florida Cord Marked, Keith Incised, Columbia Incised, Columbia Utility, and Carrabelle Punctate. The Weeden Island, Keith, and Carrabelle types indicate that the sites of Ceramic Complex C are related to the Weeden Island Complex from the northwest coast of Florida (Willey and Woodbury 1942: 242-244; Willey, 1945). Willey and Woodbury assign Weeden Island Incised to the Weeden Island I period; and a type known as Wakulla Check Stamped (Willey and Woodbury 1942: 242-243) to Weeden Island II period. It should be noted that the two types plus Swift Creek Complicated Stamped occur together on a single site.

Sites of Ceramic Complex C are more common in the Chattahoochee than those of Complex B. If the Complex B sites from the area of Seale, Alabama, are not counted, the 1947 survey found six Ceramic Complex B sites in the Chattahoochee Valley. They found eleven sites of Complex C. Several of the Ceramic Complex C sites are large, and small burial mounds were associated with two of the sites (1 He 2 and 1 Br 13).

Site 1 Ho 6 is very large and probably includes several distinct village areas. It was visited in 1905 by Clarence B. Moore, who excavated a mound in the area. In his publication on the Chattahoochee area, Moore refers to this mound as "Mound near Fullmore's Upper Landing, Houston County, Alabama" (1905: 438-444). This mound, 1.5 miles southwest of Fullmore's Upper Landing, was completely excavated by Moore. The mound was small and had an average

TABLE 7

Sherd Count – Ceramic Complex C (Swift Creek-Weeden Island I)

	1Ho1	1Ho6	1He1	1He2	1He17	1He21	1He39
Carrabelle Incised	1	3					
Carrabelle Punctate		4	1		2	2	1
Columbia Incised							
Columbia Red-Utility		1					78
Comer Punctate							
Keith Incised		3	1				
Mossy Oak Simple Stamped							
Mound Field Net Marked						2	
Seale Plain	3						
Stallings Plain					3		
Swift Creek Complicated Stamped		3	1	29	40	13	1
Tucker Ridge Pinched		1					
Weeden Island Punctate							
Weeden Island Red Filmed						15	
Weeden Island Plain	64	124	35	45	69	65	
West Florida Cord Marked			1			4	
Wilson Check Stamped	8	3	4	6	3	15	1
Other			4			3	
TOTAL	76	142	47	80	117	119	81

1Br4	1Br13	1Br15	1Br25	1Ru3
		8		4
	5	1		
	3		1	
	20	169	10	
	17			
	3			
			2	
			2	
	10	7	2	
	2			
	2			
	35	41		
2	1	2		
	20	8		
26		7	40	1
28	118	243	57	5

height of 3 feet. In several places in the fill were fragments of human bone. Also in the fill were fragments of decomposed chert, ranging in size from somewhat larger than a human head to somewhat smaller. One to four chert masses were found with each single burial. Numerous other masses of this stone present in the mound may represent decayed burials. From the eastern margin of the mound continuing well toward the center was the usual deposit of earthenware—scattered sherds ranging from fragments to entire vessels. Seven restorable mortuary vessels were recovered. These vessels were of the typical Weeden Island mortuary types, far more elaborate than any of those found during the 1947 survey.

In the 1947 survey, a cross trench was cut through mound 1 He 1, which lies in the woods a short distance south of site 1 He 2. Before the 1947 excavation, this dome-shaped mound was 40 feet in diameter and 3 feet high above the surface. In its center was a previous excavation with a diameter of 5 feet and a depth of 3 feet. On June 19, 1947, the writer and two laborers made an east-west cross trench 40 feet long, 5 feet wide, and 6 feet deep, through the mound. The profile showed the mound to be lens-shaped rather than dome-shaped. The height of the mound in the center was 6 feet, 8 inches.

The mound was composed of sand and a few small lenses and charcoal. Only three small sherds were found. A single fragment of a human skull lay 24 inches below the surface, 10 feet east of the west edge of the cross-trench. A projectile point with a constricted stem, type SF, was the only artifact encountered. The mound itself rested on a layer of light-colored sterile sand mixed with brown stones and gravel.

Artifacts from Ceramic Complex C

Several artifacts were found on the surface of the sites of Ceramic Complex C. These include:

Gorget. Made of an oval-shaped slab of polished red sandstone. The edges are double beveled and a hole was drilled through the center. Artifact broken. Length, 57 mm.; width, 38 mm.; thickness, 6 mm. From site 1 Br 13.

Slate Slab. A polished slab of slate of unknown use. Rectanguloid with rounded corner. Fragmentary. Length, 62 mm.; width, 58 mm.; thickness, 30 mm. From site 1 Br 13.

Handstone. A handstone made of a pebble. Pecked and polished surfaces. Length, 100 mm.; width, 97 mm.; thickness, 30 mm.

Double-bitted Axe. Percussion-flaked axe of crude workmanship. A notch on each side. Length, 140 mm.; width, 82 mm.; thickness, 48 mm. From site 1 Br 15.

Chopper. A unifaced, percussion-flaked chopper. On the flaked surface, a portion of the original nodule is present. Length, 112 mm.; width, 90 mm.; thickness, 60 mm. From site 1 Ho 1.

Projectile Points (Figures 13, 14, and 16)

NAa - lanceolate-shaped point, six examples.

NAb - more slender lanceolate point, four examples.

NB1 - triangular-shaped point, nine examples.

NC - point with straight base and rounded sides, one example.

SAc - side-notched triangular point with straight base, one example.

SAd - side-notched point with concave base, two examples.

SBb - stemmed point wich concave base, two examples.

SC - stemmed point with round base, six examples.

SD - stemmed point with wide, round base, one example.

SE - point with a wide, expanded base, one example.

SF - point with a constricted stem, three examples.

SG - corner-notchcd point, three examples.

SH - a small stemmed point with a rounded base, one example.

SJ - a slender point with a narrow, square stem, one example.

O - a large, parallel-sided point, three examples.

E2 - blunt with a stemmed and concave base, one example.

Drills and Blades (Figure 15)

DA - cylindrical-shaped drill, six examples.

M1 - a blade with a single side notch, one example.

M3 - large, percussion-flaked, asymmetrically shaped blades, 17 examples.

M4 - large, oval-shaped, percussion-flaked, bifaced blade, one example.

CERAMIC COMPLEX D

Ceramic Complex D sites were so classified on the basis of sherds collected from the surface. A distinguishing feature of these sites, compared to Ceramic Complex C sites, is lack of Swift Creek Complicated Stamped and increasing importance of Wilson Check Stamped as a major decorated pottery type.

A great variety of pottery types were assigned to this complex: Weeden Island Punctate, Weeden Island Plain, Weeden Island Red Filmed, Carrabelle Punctate, Carrabelle Incised, Stallings Plain, Wilson Check Stamped, Mound Field Net Marked, Keith Incised, Columbia Incised, Columbia Utility, West Florida Cord Marked, Unnamed Type Nine, and Mossy Oak Simple Stamped. On some of the sites, examples of Coweta Micaceous Utility were found. As this ware is a major type associated with Historic Creek sites, the presence of these examples is probably accidental.

The sites of Ceramic Complex D appear to be related to the Weeden Island II sites from the northwest Florida coast area, where Wakulla Check Stamped is the marker type (Willey and Woodbury 1942: 243-244). Willey notes, however, that in the Weeden Island II of the northwest Florida coast, a small amount of Swift Creek Complicated Stamped (Late Variety) still remains (Willey 1949a: 101). Sites in Alabama assigned to Ceramic Complex D lack Swift Creek Complicated Stamped.

A population climax was reached in the Chattahoochee Valley during the Weeden Island II period. Forty-seven Ceramic Complex D sites were found in the 1947 survey, in contrast to the 11 sites of Ceramic Complex C. No mounds can be definitely associated with Ceramic Complex D, although temple mounds were found on sites containing Complex D pottery sherds together with sherds of the Fort Walton and Lamar complexes. It is highly probable that these mounds belonged to the latter complexes. No sites of Ceramic Complex D were excavated in the 1947 survey. It is to be noted that the majority of the sites were concentrated in Houston and Henry counties. Very few were found in Barbour or Russell counties, indicating a center of this complex in the Lower Chattahoochee Valley.

Artifacts from Ceramic Complex D

Nutting Stone. Made of a slab of ferruginous sandstone. One surface that has been pecked has a single depression. Length, 70 mm.; width, 65 mm.; thickness, 37 mm. From site 1 He 20.

Nutting Stone. Made from a slab of ferruginous sandstone. One side has cup-shaped depression. The other side is smoothed from use as a grinding stone. Length, 92 mm.; width, 74 mm.; thickness, 30 mm. From site 1 He 20.

Elbow Pipe. Made of polished mica schist. Broken both at top of bowl and end of stem. Length, 112 mm.; width, 86 mm.; thickness, 44 mm. From site 1 He 20.

Projectile Points. (Figures 13, 14, 16)

TC - triangular point with concave base, one example.

NAb - slender triangular point with round base, one example.

NB1 - triangular point with straight base, nine examples.

NB2 - a small triangular point with straight base, one example.

SA - side-notched triangular point with serrated sides, two examples.

SAb - wide, triangular point with side notches, five examples.

SAe - wide, side-notched point with rounded base, two examples.

SBa - square, stemmed point, three examples.

SBb - stemmed point with concave base, six examples.

SBc - similar to *SBb* with flaring tangs, twelve examples.

SC - rounded, stemmed point, six examples.

SD - point with a wide stem and rounded base, seven examples.

SE - expanded base, stemmed point, two examples.

SF - constricted stemmed point, seven examples.

SG - corner-notched point, five examples.

SK - diamond-shaped point, one example.

O - parallel-sided, straight-based point, one example.

El - square-stemmed, blunt with concave base, one example.

Drills, Blades, and Other Chipped Stone Artifacts.

DA - cylindrical drill, three examples.

TABLE 8

Sherds from Ceramic Complex D (Weeden Island II)

	1Ho2	1Ho2A	1Ho3	1Ho4	1Ho7	1Ho8	1Ho9
Carabelle Incised	1					1	
Carabelle Punctate							
Columbia Incised							
Columbia Red-Utility	14		222				9
Coweta Micaceous Utility							
Keith Incised							
Mossy Oak Simple Stamped	2						
Mound Field Net Marked							
Stallings Plain	5						
Unnamed Type 9							
Weeden Island Incised	1						
Weeden Island Plain		21		17			9
Weeden Island Red Filmed							
West Florida Cord Marked (Late Variety)							1
Wilson Check Stamped	4	9	4	9	1	1	3
Other	14	1	14	1	2		
TOTAL	41	31	240	27	3	2	22

1Ho11	1Ho12	1Ho13	1Ho14	1Ho17	1Ho20	1Ho21	1He3	1He4
1	2				2		1	
					1			
	42			28	218		43	
					1			
	1				7			
						3		
	1							
8	22	2			1	33	104	1
							2	
	1							
3	48	2	1	2	2	13	9	2
	14	1		99	35		41	
12	131	5	1	129	267	49	200	3

TABLE 8 (Continued)

Sherds from Ceramic Complex D (Weeden Island II)

	1He5	1He6	1He7	1He8	1He10	1He12	1He14
Carrabelle Incised							
Carrabelle Punctate							1
Columbia Incised							
Columbia Red-Utility					20		2
Coweta Mica-ceous Utility							
Keith Incised							
Mossy Oak Simple Stamped							
Mound Field Net Marked	1						
Stallings Plain				1			
Unnamed Type 9							
Weeden Island Incised							
Weeden Island Plain	4	1	4	6		15	5
Weeden Island Red Filmed		1		2			
West Florida Cord Marked (Late Variety)					1		1
Wilson Check Stamped			7	3		2	1
Other							
TOTAL	5	2	11	12	21	17	10

1He18	1He20	1He24	1He26	1He27	1He29	1He30	1He32	1He36
			1		1	1		
	7				1		2	
	1							
5	17	29		68	128	79		121
					1			
					2			
	1				2			
				2				
	1							
					1			
4	9			16			90	2
					1			
	2		2	1			1	
1	2	3		3		1	21	1
3	34	25	24	3	3	2	3	
13	74	57	27	93	140	83	117	124

TABLE 8 (Continued)

Sherds from Ceramic Complex D (Weeden Island II)

	1He41	1He42	1He43	1He46	1He47	1Br8	1Br9
Carrabelle Incised							
Carrabelle Punctate							
Columbia Incised		1	1			1	
Columbia Red-Utility		80				63	38
Coweta Micaceous Utility							
Keith Incised							
Mossy Oak Simple Stamped							
Mound Field Net Marked							
Stallings Plain		1					
Unnamed Type 9							
Weeden Island Incised							
Weeden Island Plain	10	1	39	2	93		
Weeden Island Red Filmed		1					
West Florida Cord Marked (Late Variety)							
Wilson Check Stamped	7	5	7	4	3	2	2
Other		3		42		24	14
TOTAL	17	92	47	48	96	90	54

1Br17	1Br27	1Br28	1Br29	1Ru26
		1	1	
88		5		
				4
	1	12	2	13
		2		
	1			
2	2			
6	1	1		1
96	5	21	3	18

DB1 - drill made from a square-stemmed point, one example.

M3 - large, asymmetrical, bifaced blade, six examples.

M4 - large, oval bifaced blade, five examples.

M5 - large, bifaced, triangular blade, one example.

CERAMIC COMPLEXES E, F, AND G

The sites of Ceramic Complexes E, F, and G in the Chattahoochee Valley have all been defined by the 1947 survey. Since the separation was made on the basis of sherds collected from the surface, differences between these complexes may be more apparent than real. Nearly half the sites were assigned to Complex F, although they also contained pottery that is characteristic of Complexes E and G, indicating that there may not have been such a thing as a site with only Complex E or F. In spite of the strong possibility that in the Chattahoochee, Complexes E and F may have blended into a mixed cultural tradition, these complexes are clearly separated in other regions of the Southeast.

Ceramic Complex E is characterized by the presence of Fort Walton Incised as a major pottery type, while Complex G is distinguished by such types as Lamar Complicated Stamped, Lamar Incised, Abercrombie Shell, and Abercrombie Plain. Ceramic Complex F includes both the Fort Walton and Lamar types. In addition, all three complexes have the following types: Chattahoochee Brushed, Sub-type A; Ocmulgee Fields, Sub-type B; Lake Jackson Plain; Coweta Micaceous Utility; and Mossy Oak Simple Stamped.

In the 1947 survey, 11 sites characterized by Ceramic Complex E were found, 16 sites characterized by Ceramic Complex F, and 15 sites characterized by Ceramic Complex G. Ceramic Complex E sites tended slightly to be concentrated in Houston and Henry counties; Complex G sites tended to be concentrated in Henry, Barbour, and Russell counties, indicating that the center of Ceramic Complex E was to the south and that of Ceramic Complex G, to the north. There is nothing surprising about this distribution, as Ceramic Complex F is related to the Fort Walton Complex of the northeast coast of Florida and Complex G, to the Lamar Complex of central Georgia (Willey and Woodbury 1942: 250-251; Kelly 1938: 46-51).

The sites with Ceramic Complexes E, F, and G belong to the time period generally classified as "Protohistoric," the time when European trade goods, but not necessarily European traders, were entering the area. It should be noted that none of the sites assigned to these complexes, which are single-component, are recorded on any of the maps or historic documents. All the villages that are so described date after the Yamassee War of 1715 and have on them sherds indicative of Ceramic Complex H or I. Therefore, it is doubtful that any of the pure Complex E, F, and G sites date after A.D. 1715, nor any earlier than the time of the DeSoto Expedition of A.D. 1539-1543.

To obtain additional information on the time duration of Complexes E, F, and G, it is necessary to consider related sites in other areas. Complex G is most closely related to the Lamar focus of central Georgia. The pre-1715 date of the Lamar focus is supported by evidence from the Macon Trading Post excavated by Kelly at the Ocmulgee Old Fields in central Georgia. According to Kelly, "The identification of glass beads, pipes, guns, copper or brass hawk bells, a brass weight marked with a date of 1712 found in Ocmulgee Fields some years ago by a visitor, and a Spanish coin, together indicate an approximate date for the Macon Trading Post of between 1675 and 1718" (Kelly 1938: 55). It should be noted that at the trading post the diagnostic pottery type was Ocmulgee Fields Incised and that Lamar Complicated Stamped pottery was no longer made (Kelly 1938: Plate 12b).

Though difficult to establish, the beginning date of the Lamar Complex of Georgia probably is no older than the DeSoto Expedition. Charles H. Fairbanks hypothesizes that during the late prehistoric period in Georgia, the Mississippian Pattern was first manifested by the Etowah cultural complex; then by the Savannah and the Lamar (Fairbanks 1950: 148-150). Fairbanks states that in the Tarver site, north of Macon, trade articles were associated with Lamar Complicated Stamped sherds, but that Ocmulgee Fields pottery was also present and that the Lamar pottery appeared to be of the last type made. He concludes that in central Georgia the Lamar focus first appeared during the early part of the seventeenth century and gave way to the Ocmulgee Fields focus by about 1690, while in north Georgia it continued into the late eighteenth century among the historic Cherokee.

Joseph Caldwell adds this information on the date of the Lamar focus of Georgia:

> It is significant that Spanish pottery of the 'olive jar' type . . . is found on the Georgia Coast at sites which exhibit the generalized Irene-Lamar complexes instead of the later pottery of Central Georgia. Spanish pottery has been noted on the surface collections from sites of this type on Creighton Island and in archaeological sites of the Historic Huspaw, at the site of Fort King George at Darien . . . This properly reflects the lapse of time between the historic penetrations of the coastal regions and the hinterland (Caldwell 1941: 3).

Hale G. Smith states that the Fort King George site was occupied in 1715 and that ceramics there can be classified as Late Lamar (Smith 1948: 314-315).

Ceramic Complex G from the Chattahoochee Valley of Alabama closely resembles the Fort Walton focus of the northwest Florida coast. Willey and Woodbury assign the Fort Walton complex of the northwest Florida coast to the time period following Weeden Island II, when the Mississippian influences were entering the region and European trade goods were beginning to filter into the area (Willey and Woodbury 1942: 254). They state that some of the Fort Walton sites yielded such artifacts as European metal and glass beads; others did not. John Griffin notes with regard to recent work in Florida that the Weeden Island

TABLE 9

Sherds from Ceramic Complex E (Fort Walton)

Pottery Types	Ho2	Ho12	Ho15	Ho16	He9	He21	He37	Br12	Br18	Ru1	Ru2
Abercrombie Plain											9
Abercrombie Shell Tempered											28
Chattahoochee Brushed								1			36
Chattahoochee Brushed Sub-Type A											
Coweta Micaceous Utility				3	35			1		1	17
Fort Walton Incised	1	1	1	1	1	3	2	2	1	1	
Lake Jackson Plain			10			73	53	1	4	15	
Lamar Complicated Stamped											
Mossy Oak Simple Stamped				16					1	13	
Ocmulgee Fields Incised										1	1
Ocmulgee Fields Incised Sub-Type A										1	17
Ocmulgee Fields Incised Sub-Type B											
Rocker Stamped Ware											1
Unnamed Type 9					1						
Weeden Island Plain											
Other	40	127				116			3	1	1
TOTAL	41	128	11	20	37	192	55	5	9	33	110

TABLE 10

Sherds From Ceramic Complex F (Fort Walton-Lamar Sites)

	Ho3	Ho20	He3	He11	He22	He24	He27	Br8	Br9	Br11	Br14	Br15	Br16	Br22	Br25	Woolfolks, Georgia
Chattahoochee Brushed			3					2	5		1				2	
Chattahoochee Brushed Sub-Type A					1											1
Ocmulgee Fields Incised	1		3		1	1	1	2	3			1		1	8	18
Ocmulgee Fields Incised Sub-Type A			8	1		1		14	3			2			12	2
Ocmulgee Fields Incised Type B			1			1		3							1	1
Unnamed Type 9				1												
Lamar Complicated Stamped	1	1	18	4	1		1	1		1			2	2		3
Fort Walton Incised	7	4	5	1	1	20	1	1	1	1	7	1	2	3	2	3
Abercrombie Shell Tempered		31				1		1	1			1			29	106

TABLE 10 (Continued)

	Ho3	Ho20	He3	He11	He22	He24	He27	Br8	Br9	Br11	Br14	Br15	Br16	Br22	Br25	Woolfolks, Georgia
Abercrombie Incised									1							
Abercrombie Plain														43		23
Lake Jackson Plain				25	32						80				10	69
Coweta Micaceous Utility	1		3		1	1					1		5	21		
Mossy Oak Simple Stamped	1			1	3				5	9				12		1
Weeden Island Plain	2															
Rocker Stamped Ware	5															
Other	222	224	159			35	104	66	40			199	6		6	
Total	240	260	200	33	40	60	107	90	59	11	89	204	15	82	70	227

TABLE 11

Sherds From Ceramic Complex G (Lamar Sites)

	Ho17	He19	He25	He34	He40	He46	Br6	Br17	Ru6	Ru7	Ru11	Ru19A	Ru21	Ru22
Abercrombie Plain		1		3										57
Abercrombie Shell Tempered				2					2	2	2		5	
Chattahoochee Brushed											2	6		
Chattahoochee Brushed Sub-Type A														
Coweta Micaceous Utility		3		2	2						9	26		
Fort Walton Incised														
Lake Jackson Plain	28	10		7	28					3	32	12		15
Lamar Complicated Stamped	2		1	3		42	1	6	2		3	1		
Mossy Oak Simple Stamped														

TABLE 11 (Continued)

	Ho17	He19	He25	He34	He40	He46	Br6	Br17	Ru6	Ru7	Ru11	Ru19A	Ru21	Ru22
Ocmulgee Fields Incised		3										1		
Ocmulgee Fields Incised Sub-Type A		6		7	1				1		7	1		
Ocmulgee Fields Incised Sub-Type B														
Rocker Stamped Ware														
Unnamed Type 9														
Weeden Island Plain														
Other	99		113			6	4	90	36			4	213	132
Total	129	23	114	24	31	48	5	96	41	5	55	51	218	204

period ended at approximately A.D. 1450 and the Fort Walton focus extended from this date until A.D. 1600 or later (Griffin 1950: 111).

Hale G. Smith has defined a Leon-Jefferson period in the Tallahassee area of Florida that succeeded the Fort Walton (Smith 1948: 315-319). He dates this period from 1650-1750, a time coeval with the Spanish missions of the area. At this time Fort Walton Incised continued there, and check-stamped sherds reappeared. A type that he names Mission Red Filmed was evidently related to the Kasita Red Filmed pottery associated with the Ocmulgee Fields focus (Ceramic Complex H, in the Alabama part of the Chattahoochee Valley). This relationship indicates at least an overlap in time between the Leon-Jefferson and the Ocmulgee Fields complexes.

The limited amount of testing at sites of Ceramic Complexes E, F, and G by the 1947 survey was not sufficient to determine whether all three of these complexes were coeval or overlapping in time. Nor could it be proven on the basis of the surface collections by the 1947 survey that these complexes were unrelated.

A common feature of the larger sites of Complexes E, F, and G was the temple mound. Single temple mounds were associated with sites 1 Ho 3, 1 Br 14, 1 Br 15, and 1 Ru 10; two mounds were found at site 1 Br 8.

During the 1947 survey a limited amount of testing was done in site 1 Ho 3, a large village site located a short distance south of Columbia, Alabama. Here is found the mound described by Clarence B. Moore as "Mound below Columbia Landing." This mound surmounts a high cliff overlooking the Chattahoochee River, and is surrounded on the south and west by the village site. At the time of Moore's visit in 1906, this temple mound was 8.5 feet high and 138 feet long. It measured 57 feet across the north end, 68 feet through the center, and 88 feet across the south end. The flat summit plateau was 93 feet long with a maximum width of 54 feet. Moore sank several test pits into the mound but was unable to find any burials (Moore 1905: 445-446). In the 1947 survey, two 5-by-5-foot test pits were cut in the village area to the south of the mound. The occupation level was 1.5 feet thick.

Pottery stratigraphy in 1 Ho 3 indicates Fort Walton Incised, Lamar Complicated Stamped, and Wilson Check Stamped were made at the same time.

Site 1 Br 8 was one of the largest villages in the area that had a Lamar, Fort Walton, or Complex F occupation. A Complex D (Weeden II) was also present here. This site lies about 3 miles north of Louisville, Alabama, or about 22 miles west of the Chattahoochee River. It is well known to the local inhabitants and was previously described by Miss Anne Kendrick Walker in her book *Backtracking in Barbour County* (1941).

Two temple mounds are present at this site. Mound 1 Br 3 lies about 40 paces west of Mound 1 Br 4, and is near the center of the village. In 1947, it stood 11 feet high, 60 feet in diameter, and 270 feet in circumference. Cultivation has reduced this mound to a dome shape.

Mound 1 Br 4 was approximately 8 feet high and 329 feet in circumference in

TABLE 12
Pottery Count of Two Stratigraphic Trenches at Site 1 Ho 3

	Fort Walton Incised	Wilson Check Stamped	Lamar Complicated Stamped	Coweta Micaceous Utility	Lake Jackson Plain	Ocmulgee Fields Incised	Ocmulgee Fields Incised (Sub-Type A)	Carrabelle Incised	Total
Strat. No. I									
0-6"	8.9% 2	4% 1			79% 19	4% 1		4% 1	24
6"-1'	3% 2				97% 69				71
1'-1'6"		3% 1		3% 1	91% 31		3% 1		34
Strat. No. 2									
0-6"					100% 4				4
6"-1'	3.4% 3	2.3% 2	1% 1		93% 84				90
1'-1'6"					100% 15				15

1947. Probably it was originally of flat-topped, pyramid shape, but it has now been reduced by cultivation to a dome shape. Local inhabitants report that this mound formerly stood 15 feet high. Near the mound a skull was supposed to have been plowed up.

Another well-known site with a temple mound is site 1 Br 15, located south of Francis Bend of the Chattahoochee and a few miles north of Eufaula, in Barbour County. Clarence B. Moore visited the site in 1906, and sank several test pits in Mound 1 Br 2 with negative results (Moore 1906: 448). The mound parallels a high cliff of the Chattahoochee. The river side of the mound has been washed away, revealing red clay fill that rests on an old humus line plainly visible in the eroded banks. A few sherds were seen in 1947 in both the humus line and mound fill. The mound was 7 feet high and 25 feet wide at the base in 1947. The flat top of the mound was 50 feet long and 8 feet wide. At least one half of the mound has been washed away by the river.

About two miles north of Upper Francis Landing on the Shorter Plantation is another large pyramidal mound. The site (1 Br 14) was not visited by C. B. Moore, although he referred to the mound in his report of 1905 (Moore 1905: 448). The mound (1 Br 3) lies buried under deep trees and shrubs near the north side of a dry lake bed. There has been very little damage to the mound, which is oblong-shaped with a slightly rounded top. If the mound were cleared, it would probably be an elongated, flat-topped pyramid. The height is about 13 feet and the circumference 360 feet. The village area to the north appeared to be rich, and sherds were plentiful on the surface.

The most intensive testing in the 1947 survey was at Abercrombie Village, 1 Ru 10. Although this site contains Ceramic Complex E, it will be described separately in this report because of the large amount of artifacts and sherds found in the test trenches.

Artifacts from Ceramic Complexes E, F, and G

Shell Scraper. Fragment of a clam shell scraper. Length, 71 mm.; width, 37 mm.; thickness, 3 mm. From 1 Br 8.

Shell Scraper. Fragment of a clam shell scraper. Length, 47 mm.; width, 35 mm.; thickness, 3 mm. From site 1 Ru 21.

Shell Pendant. Teardrop-shaped, columella pendant. Length, 40 mm.; width, 10 mm.; thickness 8 mm. From site 1 He 3.

Stone Hoe. Rectangular hoe made of chlorite schist. The bit is polished but the side is fairly rough. One side is broken. Length, 115 mm.; width, 60 mm.; thickness, 16 mm.

Projectile Points (Figures 13, 14, and 16)

TC - triangular point with concave base, one example.

NAa - triangular point with high, rounded base, one example.

NB1 - triangular point with straight base, one example.

NB2 - very small, triangular point with straight base, one example.

NC - small, slender, triangular point, straight base, one example.

SBa - square-stemmed point, two examples.

SBb - square-stemmed point with concave base, two examples.

SBc - stemmed point, concave base, flaring shoulders, four examples.

SC - stemmed point with rounded base, five examples.

SD - stemmed point with wide, rounded base, three examples.

SF - constricted-stemmed point, one example.

SG - corner-notched point, two examples.

SG1 - corner-notched point, with tangs extending to base, one example.

SH - very small, rounded-stemmed point, one example.

E1 - square-stemmed, blunt point, three examples.

Drills and Blades (Figure 15)

DA - cylindrical drill, one example

M2 - small, square blade about size of postage stamp, one example.

M3 - large, asymmetrical, bifaced blades, four examples.

M7 - drill made by reworking a slender, stemmed point, one example.

M8 - small, snub-nosed end scraper, one example.

Notes on Projectile Points and Other Chipped Stone Artifacts

Probably most of the stemmed points and the corner-notched points do not actually belong to Ceramic Complexes E, F, and G. These types are generally characteristic of the older cultural horizons in the Chattahoochee Valley. Their presence is probably explained by the fact that many of the sites with either Ceramic Complex E, F, or G are multicomponent—the larger projectile points probably belong to an earlier horizon. It should be noted that at Abercrombie Village, 1 Ru 10, none of the larger stemmed and corner-notched points was uncovered in the test trenches. The general lack of chipped stone work associated with the Fort Walton (Ceramic Complex E) was noted by Ripley Bullen in the Chattahoochee Valley of Florida. He tested eight undisturbed Fort Walton sites and found very few chips (Ripley P. Bullen, *personal communication*, December 31, 1948).

CERAMIC COMPLEXES H AND I

The sites that have been assigned to Ceramic Complexes H and I were classified on the basis of surface collections by the 1947 survey. Their ceramic content was very similar, although Ocmulgee Fields Incised pottery was absent from Ceramic Complex I. It is quite possible that this difference in ceramic content may be attributed to the fact that surface collections in the area were rarely large enough to include all of the types that may have been present at the site. On the basis of its lack of Ocmulgee Fields Incised, Ceramic Complex I would appear to be later, dating from the time when native-made ceramics were being abandoned in favor of European trade goods. However, several of the sites that apparently were visited by Benjamin Hawkins in 1799 appear to have belonged to Ceramic Complex H (for example, site 1 Ru 11, previously mentioned in this report as a possible site of Coweta Tallahassee).

The pottery types common to Ceramic Complexes H and I include Coweta Micaceous Utility, Lake Jackson Plain, Chattahoochee Brushed, Sub-Type A, Kasita Red Filmed, and Unnamed Type 9. On several of the Ceramic Complex H sites, examples of Lamar Complicated Stamped, Abercrombie Shell Tempered, Fort Walton Incised, and Type 10 (an unnamed, rocker stamped ware) were found. As most of these types belong to Ceramic Complex F (Fort Walton-Lamar), it is highly probable that sherds of these types on Complex H sites are a result of multi-occupation.

Site 1 Ru 12 has a ceramic content divergent from the content of Complexes H and I. Differences are manifested in the very high proportion of Lake Jackson Plain and the lack of typical Historic Creek sherds such as Coweta Micaceous Utility and Chattahoochee Brushed.

Few artifacts were found on the Historic Creek sites, and as many of these sites displayed clear evidence of earlier occupations, it is possible that most of these artifacts were part of the earlier complexes. Artifacts that accompanied the burials in the cemetery at Abercrombie Village, however, very probably are Historic Creek. Their descriptions are found in this report in the section on Abercrombie Village, site 1 Ru 10 (1 Ru 61).

The following artifacts were found on sites that did not show clear evidence of earlier occupations. The celts and fragment of atlatl weight or bannerstone described below may not be Historic Creek artifacts.

Atlatl Weight. Fragment of a bannerstone made of polished red shale. Through the center was a longitudinal hole. Length, 40 mm.; width, 28 mm.; thickness, 16 mm. From 1 Br 3.

Celt. Petaloid celt made of diorite. Pecked surface with a smooth poll and polished blade. Length, 150 mm.; width, 58 mm.; thickness, 34 mm. From 1 Ru 12.

Celt. Petaloid celt made of a silicate rock. Square poll with smoothed, unpolished surface. Broken lengthwise and crosswise. Length, 40 mm.; width, 45 mm.; thickness, 28 mm. From site 1 Ru 23.

Columella Pendant. Cone-shaped pendant made of a shell columella. Smoothed surface. Broken lengthwise. Length, 31 mm.; width, 11 mm.; thickness, 8 mm. From 1 He 32.

Copper Bead. Tubular bead made from a wrapped piece of sheet copper. Length, 35 mm.; diameter, 15 mm.; thickness, 1 mm. From 1 Ru 12.

Glass Fragment. A piece of green-colored glass. Surface is highly patinated. On outer surface is an embossed "S." From 1 Ru 18.

Hoe Blade. Rough rectanguloid hoe made from a slab of chlorite schist. The front face is smoothed but the reverse shows large percussion flakes. Length, 85 mm.; width, 76 mm.; thickness, 13 mm. From 1 He 33.

Mortar. Made from a rectangular slab of ferruginous sandstone. The circular concave depression was 4 mm. deep, 48 mm. long, 105 mm. thick.

Pipe. Stem and portion of bowl from an English claypipe. Has a ridged or corrugated surface with a smooth finish. Fragment is 27 mm. long, 25 mm. in width. From site 1 He 25.

Pipestem. Fragment of a cylindrical pipestem made of a gray pottery. Smooth surface. Length, 23 mm.; diameter, 23 mm.; thickness, 10 mm. From 1 He 33.

Projectile Points. (It is doubtful that the stemmed, side-notched points, and corner-notched points included in the following description belong to the period of Historic Creek occupation. Points are illustrated in Figures 13, 14, and 16.)

NAa - leaf-shaped point, one example.

NB1 - large, triangular point, one example.

NBc - small, triangular point with slightly flanged base, one example.

SAb - side-notched, triangular point, one example.

SAd - side-notched, triangular point with rounded base, one example.

SBa - square-stemmed point, one example.

SBb - square-stemmed point with concave base, five examples.

SBc - square-stemmed point with tanged shoulders and concave base, six examples.

SC - round-based, square-stemmed point, four examples.

SD - similar to SC but with a very wide stem, four examples.

SF - constricted-stemmed point, three examples.

SF1 - similar to SF but has a more rounded stem, three examples.

SG - corner-notched point, six examples.

SH - very small point with a square stem, one example.

SK - a point with a triangular base, one example.

Blades, Drills, and Other Chipped Stone Artifacts (Figure 15)

Drill DA - cylindrical drill, one example.

Scraper L1 - small, rectangular artifact with a notch in base, one example.

Blade M1 - blade with a single side shoulder, one example.

Scraper M2 - small rectangular scraper, two examples.

Blade M3 - large, asymmetrical blade, three examples.

Ceramic Types in the Chattahoochee Valley

ABERCROMBIE INCISED (Figure 12)

Paste:

Temper: Grit, shell, grit and shell.

TABLE 13

Sherd Count – Ceramic Complex H (Ocmulgee Fields)

	He13	He19	He20	He28	He33	He34	He40	Br10	Br28	Ru2
Abercrombie Plain		1				3				9
Abercrombie Shell Tempered						2				28
Chattahoochee Brushed	9				13					36
Chattahoochee Brushed Sub-Type A			1	2			1	1	1	
Coweta Micaceous Utility		3			13	2	2		1	17
Dowel Marked Sherds										
Fort Walton Incised										1
Kasita Red Filmed										
Lake Jackson Plain	338	10		39	94	7				
Lamar Complicated Stamped						3				
Ocmulgee Fields Incised		3								
Ocmulgee Fields Incised Type A	15	6			2	7				17
Rocker Stamped										
Unnamed Type 9					1					
Unnamed Type 10										1
Other	1		71				28		4	1
Total	363	23	72	41	123	24	31	1	6	110

TABLE 13 (Continued)

Sherd Count – Ceramic Complex H (Ocmulgee Fields)

	Ru4	Ru6	Ru8	Ru11	Ru12	Ru18	Ru19A	Ru20	Ru23
Abercrombie Plain	1							27	
Abercrombie Shell Tempered		2		2				4	
Chattahoochee Brushed				2		40	6	27	9
Chattahoochee Brushed Sub-Type A								1	
Coweta Micaceous Utility	1		21	9		18	26	16	1
Dowel Marked Sherds									
Fort Walton Incised									
Kasita Red Filmed								2	14
Lake Jackson Plain	4		16	32	894	93	12	15	4
Lamar Complicated Stamped		2		3			1		
Ocmulgee Fields Incised			1			1	1	4	14
Ocmulgee Fields Incised Type A		1	5	7	1	1	1		
Rocker Stamped									
Unnamed Type 9									
Unnamed Type 10									
Other		36					4		
Total	6	41	43	55	895	153	51	96	42

TABLE 14

Sherd Count – Ceramic Complex I (Late Historic Creek)

	Ho5	Ho12	Ho17	He15	He16	He25	He26	He31	He32	He38	Br1	Br3	Br4
Chattahoochee Brushed	5	1	9	40	2		24	30	3	3	2	5	7
Chattahoochee Brushed Sub-Type A													
Coweta Micaceous Utility		10	4		3	28							9
Kasita Red Filmed					3	38							
Lake Jackson Plain				1	1	57	22	27					8
Ocmulgee Fields Incised													
Unnamed Type 9													
Other		117	116			1			114	34			2
Total	5	128	129	41	9	124	46	57	117	37	2	5	26

TABLE 14 (Continued)

Sherd Count – Ceramic Complex I (Late Historic Creek)

	Br5	Br6	Br7	Br10	Br21	Br23	Br24	Br30	Br31	Ru5	Ru9	Ru21	Ru22
Chattahoochee Brushed	8	3	5		13	12	5	70	15		14	59	70
Chattahoochee Brushed Sub-Type A					1							3	
Coweta Micaceous Utility	12		6	9	1	6	2		2	5	20	68	5
Kasita Red Filmed	1											1	
Lake Jackson Plain					1	1	13	80	12	5	27	82	57
Ocmulgee Fields Incised													
Unnamed Type 9						1							
Other		2		16								5	72
Total	21	5	11	25	16	20	20	150	29	10	61	218	204

Texture: Same as Ocmulgee Fields Incised.

Hardness: Same as Ocmulgee Fields Incised.

Color: Interior and exterior are both burnished black.

Surface Finish:

Surfaces are smoothed and polished to reflection. Burnishing and smoothing marks visible. Surfaces display a fairly thick filming or float, which has a tendency to scale off. A large amount of mica is present in the float.

Decoration:

Technique: The majority have incised lines that are much neater than those on Ocmulgee Fields Incised and not so wide as those on Lamar Bold Incised. A minority type has engraved designs that are not so carefully executed as engraved designs found on pottery from Moundville, Alabama.

Design: The incised and narrow-trailed sherds have the same design elements as the local Ocmulgee Fields Incised varieties. Engraved designs include combinations of terraced lines and concentric triangles, bands of encircling lines connected by series of diagonal or vertical lines, and other rectangular elements.

Distribution: Interior and exterior of rim areas. Bands on upper exterior of vessels.

Form:

Same as Ocmulgee Fields Incised except a greater number of notched rim fillets and sharply bent L-shaped rims.

Usual Range of Type:

In the Chattahoochee it is found frequently on site 1 Ru 10, Russell County, Alabama, and across the river at the Woolfolks site in Georgia. Sporadic examples are found on other mixed Lamar-Fort Walton sites. It bears a close resemblance to a series from around Montgomery, Alabama, collected by Peter Brannon, although the sherds in the collection he gave to the Ceramic Repository, University of Michigan, do not show engraving or such a high degree of polishing. Many of the examples from site 1 Ru 10 are similar to the type known as Moundville Black Filmed; however, Moundville Black Filmed is predominantly shell tempered, while the sherds from the Chattahoochee are grit and shell tempered. A single example of Abercrombie Incised was found on a mound site at the junction of the Tallapoosa and Coosa rivers, Alabama.

Chronological Position of Type in Range:

In the Chattahoochee Valley it occurs on sites with the Ceramic Complexes F and G (Lamar-Fort Walton). The presence of a small percentage of Ocmulgee Fields Incised on these sites probably indicates a late Lamar time period. This time would be considerably later than the date of the Moundville, Alabama site, which may have been contemporary with Weeden Island II. A Moundville, Alabama source for Abercrombie Incised is highly probable.

ABERCROMBIE PLAIN (Figure 12)

Paste:

Method of Manufacture: Coiled.

Temper: Mixed sand and grit. Occasionally contains a minute amount of medium-size shell, which is usually not visible on surface because of floating or filming. Micaceous.

Texture: Fracture, medium angular; a little pitting; compact.

Hardness: 3.5 - 4.

Color: Deep brown, orange, brown-red, various shades of buff and gray. Color may vary on a single vessel. Firing clouds. Generally have a dark core.

Surface Finish:

Modifications: Surface has been modified by a thick float or film that tends to wear off. The exterior, and occasionally the interior, has been polished to the extent that it has a low light-reflecting ability. Polishing and scraping marks may show on some specimens.

Decoration:

Technique: Confined to modifying the rim by deep notching at lip or by appliqués of the notched fillets of molded life forms. Usually vessel is undecorated.

Designs: Applied on appliqués or lip; ticking or punctating with a diamond-shaped or hollow reed instrument.

Form:

Rim: Flaring; incurved or soup-plate profile.

Lip: Great variety. Rounded, thickened, pinched, extruded, and square.

Body: Many shapes. Globular vessels with constricted necks and flaring shoulders, globular bowls with incurved rims, soup-plate forms, straight-sided vessels. Vary greatly in size.

Thickness: Rims 6 - 13 mm.; body 4 - 7 mm.

Usual Range of Type:

Chattahoochee Valley of Georgia and Alabama.

Chronological Position of Type in Range:

Part of the Ceramic Complexes F and G (Fort Walton-Lamar)

Remarks:

Closely related to Abercrombie Incised, lacking only the decoration. Includes some of the most finely made pottery in the Chattahoochee Valley. Some of it is as fine and symmetrical as wheel-turned pottery.

Variant A (Abercrombie Shell Tempered)

This ware is similar to that described above except that it is shell tempered. Occasionally widely spaced, cross-hatched lines are incised below lip on exterior.

Variant B (Abercrombie Black Filmed)

Similar to Abercrombie Plain except that it has a highly polished, burnished exterior. Rare type except at sites such as 1 Ru 10 and Woolfolks Mound.

COLUMBIA INCISED

Description:

This type appears to be a variety of Carrabelle or Keith Incised. The paste is similar to that of Lake Jackson Plain. The incising is broader, deeper, and rougher than that on Carrabelle Incised. Design elements include chevrons, diagonals, and vertical lines pendant from an encircling rim line. Forms include the shallow bowl with flaring rim, with the design placed on the inside, and bowls with constricted or extrusive notched rims.

Associations:

Associated with the Weeden Island horizon (Ceramic Complex D) in the Chattahoochee.

COLUMBIA UTILITY (Figure 11)

Paste:

Method of Manufacture: Coiled, coiling marks show.

Temper: Large grains of sand, much larger than on Weeden Island Plain. Protrudes on surface and is plainly visible.

Texture: Medium to rough angular texture – diagonal and horizontal laminations. Sandy.

Hardness: 3.5 - 4.5

Color: The most common color is a reddish buff. Sometimes it is gray. Firing clouds. Cross section may be either dark gray or light gray.

Surface Finish:

Smoothed. Contrasts with the deeply pitted and sandy surface of Seale Plain or the very smooth surface of Weeden Island Plain.

Decorations:

Notching and occasionally a fine incised line along rim.

Form:

Rim: Great variety of rims: thickened; flaring outward or inward; rims bent at sharp angles to shoulders.

Body: Vessels display constricted necks, flaring rims, and hemispheroid bowls.

Thickness: Rims 5 - 17 mm.; vessel walls 5 - 15 mm.

Appendages: Large variety of loop handles with noded or bifurcated noded tops; strap handles (rare) long, flat or rounded nodes; idented fillets and appliqués; circular dipper handles; modeled life forms such as fish and animal heads applied to vessels.

Usual Range of Type:

Chattahoochee Valley, Alabama.

Chronological Position of Type in Range:

Associated with Ceramic Complexes C, D, E, F, G (Swift Creek, Weeden Island I, Weeden Island II, Fort Walton, and Lamar).

Remarks:

This type can be distinguished from Seale Plain, which has a sandy surface and is generally much thicker. Weeden Island Plain has a finer temper and a smooth polished surface. The Chattahoochee variant of Lake Jackson Plain has coarser temper and a heavily floated, crackled surface.

COWETA MICACEOUS UTILITY (Figure 10)

Paste:

Method of Manufacture: Probably coiled.

Temper: Small to moderate amounts of fine grains of sand or grit. The clay contains a large amount of mica that is plainly visible on the surface.

Texture: Smooth to medium-rough, slightly pitted, sandy, even textured.

Hardness: 2 - 3.5. Some examples are very soft and easily broken with the fingers; others are considerably harder.

Color: The most common color is a silvery gray, varying to buff and red. Firing clouds present. Cross section often dark gray.

Surface Finish:

Well smoothed on most specimens but on a few the smoothing has produced scratches as a result of dragging the protruding temper.

Type Name:

Seale Plain (Figure 9).

Paste:

Method of Manufacture: Probably coiled.

Temper: Fine to large grains of sand; moderate to abundant amount.

Texture: Rough, pitted, sandy.

Hardness: 3.5 - 5.

Color: Yellowish-buff surface, dark cross section.

Surface Finish:

Rough surface caused by projecting coarse temper; pitting, scraping marks on interior; poorly smoothed; sandy.

Form:

Rim: Flaring.

Lip: Rounded or pinched.

Body: Unknown but apparently a conoidal vessel with flaring rim and constricted shoulders.

Thickness: 4 - 8 mm.

Usual Range of Type:

Chattahoochee Valley of Alabama.

Chronological Position of Type in Range:

Associated with the Ceramic Complex A (Deptford) and Complex B (Swift Creek), lingers in a small percentage in Complex C (Swift Creek-Weeden Island I).

Remarks:

Though difficult to describe, the differences in the surface appearance of Seale Plain and Lake Jackson Plain wares are striking.

WILSON CHECK STAMPED (Figure 8).

Paste:

Method of Manufacture: Coiled. Coil fractures are occasionally present.

Temper: Fine to medium-sized quartz sand and grit.

Texture: Medium to angular; occasionally gritty.

Hardness: 4.

Color: Varies from light reddish-buff to gray; occasionally, smudging or firing clouds are evident. It may have a gray to dark gray cross section.

Surface Finish:

Modifications: Interior surface smoothed and sandy; a few have a well smoothed interior.

Decoration:

Technique: Check stamping sometimes carelessly overlapped. The lands may be applied both parallel and oblique to the rim.

Design: Considerable variation in design application techniques; however, seriation analysis of collections from the Alabama sites failed to reveal any significant chronological or statistical differences between variations. All varieties may be found on the surface of a single site. One variety has finely executed check impressions with parallel lands and resembles Wakulla Check Stamped (Willey and Woodbury 1942: 243-244, Plate XII, Figure 1c). Other varieties have much larger and more poorly executed impressions. Some were so poorly stamped that certain areas have no impressions. None of the impressions is so deep as Deptford Bold Check Stamped impressions. Several of the sherds resemble Deptford Linear Check Stamped, but differ in that the impressions are much smaller.

Modifications: A single example had an encircling incised line below the rim.

Form:

Rim: Flaring rims and incurved rims.

Lip: Tapered, beveled, or folded. The base of the fold may be rounded or extruded. Both incurving and outcurving folds are present. The fold may have a scalloped base.

Body: Globular or deep with conoidal base.

Thickness: Body 5 - 10 mm.; rim 4 - 6 mm. Fold varies from 2 to 6 mm. in width.

Appendages: One example has a tetrapodal support.

Usual Range of Type:

Wilson Check Stamped pottery is found throughout the Chattahoochee Valley of Alabama. It is closely related to Wakulla Check Stamped from the northwest Florida coast (Willey and Woodbury: 243-244, Plate XII, Figure 1c).

Chronological Position of Type in Range:

Belongs to the Weeden Island Period of the northwest coast of Florida. In the Chattahoochee Valley of Alabama, it begins as Swift Creek Complicated Stamped, dies out (Ceramic Complex C – Swift Creek-Weeden Island I), then reappears as a major type (Ceramic Complex D – Weeden Island II). It disappears with the beginning of the Ceramic Complexes F (Fort Walton) and G (Lamar).

Variations:

One sherd that has the large, smoothed-over checks may be related to the Leon Check Stamped type defined by Smith (Smith 1948: 316, Plate XXXII, K). Some sherds display the large, irregular design of St. Johns type from Florida but are made of typical Weeden Island paste. A few have a soft micaceous paste.

CHAPTER 3

Summary and Conclusions

One hundred and twenty-four village and camp sites were located in Russell, Barbour, Henry, and Houston counties during the summer of 1947. In addition, 20 mounds and mound groups and one quarry were surveyed. Because of the heavy growth of timber and intensive cultivation in many areas, many sites may have been overlooked. A test pit was made in site 1 Ho 3, a cross trench was cut in Mound 1 He 1, and seven test pits were made in Abercrombie Village, site 1 Ru 10 (1 Ru 61). Several private collections in the area were studied for comparative material.

Because the bulk of the information contained in this report is based upon surface collections, all conclusions should be regarded as tentative. The occupation of a single site at different time periods by Indians of different cultures undoubtedly has caused a considerable admixture in the specimens collected from the surface. Before more definite conclusions about the occupants of the Chattahoochee Valley can be drawn, it will be necessary to excavate several type sites.

No pre-ceramic sites were located in the Chattahoochee Valley, although private collections contain projectile points similar to points from pre-ceramic sites in other areas. Among these are fluted points and the indented base stemmed points. With exception of a fluted point resembling the Clovis type the other points in this series were not found in the valley but in the highland areas near Seale, Alabama, a few miles to the west.

On the basis of the 1947 survey the oldest sites in the Chattahoochee Valley were assigned to Ceramic Complex A. These sites were very small—consisting of a scattered house or two—and scarce. Pottery was scarce, although flint chips were prominent on the surface. Diagnostic of this complex is the Deptford series of pottery.

This early cultural horizon was followed by Ceramic Complex B. The diagnostic pottery type was Swift Creek Complicated Stamped, of the middle to late varieties. Sites were still small and scarce—only six were found in the Chattahoochee Valley. Pottery and flint chips were more common in Complex B. The origin of this complex may have been in central Georgia.

Following Complex B was Ceramic Complex C, characterized by Swift Creek Complicated Stamped and a variant of Wakulla Check Stamped designated as Wilson Check Stamped. A marked increase was noted in the number of sites—11 were encountered. Small burial mounds were associated with two of the sites. The presence of Wilson Check Stamped and the Carrabelle series indicates a close relationship with the Weeden Island II complex from the northwest coast of Florida described by Willey and Woodbury.

A population climax was reached during the time period marked by Ceramic Complex D. Forty-seven sites (in contrast to 11 of the previous period) were located. The pottery type diagnostic of Ceramic Complex D was Wilson Check Stamped. Swift Creek Complicated Stamped was absent. No burial mounds could be definitely associated with this complex. The majority of the sites were concentrated in the lower part of the Chattahoochee Valley covered by the 1947 survey, while in the upper counties of Barbour and Russell the sites were rare. Many of the Complex D sites were larger than those of the earlier Complex C. Complex D was probably coeval with Weeden Island II.

The following cultural horizon, represented by Ceramic Complexes E, F, and G, occurred when diverse cultural influences were entering the region; however, because of lack of sufficient excavation in these sites it is possible that there were not actually three distinct cultural complexes in this time horizon. Complex E refers to those sites marked by Fort Walton Incised pottery, Complex F to sites with both Fort Walton Incised and Lamar Complicated Stamped.

The Fort Walton-Lamar complex occurred at a time when Middle Mississippian traits were entering the area in large numbers. Single and paired flat-topped temple mounds were associated with the larger sites. Mounds at the Lamar sites in the northern counties tended to be of the square, truncated type set within the village area, while a couple of mounds on the Fort Walton sites in the lower counties had a more elongated shape and were constructed at the edge of the village, parallel to the river bank.

A very small amount of Wilson Check Stamped pottery was found on sites that seemed to be predominantly of the Fort Walton type, indicating that there was probably no time gap between Complex C (Weeden Island II) and Complex E. On the sites that appeared to be predominantly Walton, admixture with the Lamar complex was found, and it is questionable if a "pure" Fort Walton site existed in the area covered by the 1947 survey.

It is equally questionable if a "pure" Lamar site existed in the area. Those assigned to Complex G lacked important Lamar types such as Lamar Bold Incised and, in addition, contained Fort Walton types, blackfilmed and engraved sherds that reflect the Moundville cultural tradition, other sherds similar to those from the "Burial Urn Complex" from central Alabama, and Ocmulgee Fields Incised.

Without extended excavation, it is difficult to relate Complex G sites to specific movements of Creek tribes. It cannot be determined whether sites dated from the period when the Creek were moving eastward across Alabama and Georgia or from a time after the Yamassee War of 1715, when they migrated from Georgia back into the Chattahoochee Valley. Probably there are sites of Complex B that belong to both phases in the Creek migration. The chronological difficulty is illustrated at Abercrombie Village. The presence of the Moundville type of sherds supports a theory that this village belonged to the earlier migration, while the presence of Ocmulgee Fields Incised reflects a post-Yamassee War date. There was no question of the association of Ocmulgee Fields Incised and the Moundville types of black filmed, engraved sherds at Abercrombie, for sufficient excavations were made to establish this fact.

A grand total of 47 sites of Ceramic Complex E, F, and G was located in the 1947 survey. Although the total is the same as the number of Complex D sites found, many of the sites of these three more recent complexes seem to be larger in size than those of Complex D, indicating a slight population increase.

The Historic Creek in the Chattahoochee Valley were represented by Complexes H and I. Because the classification was based upon surface collections, this separation may be more apparent than real. Complex H is marked by Ocmulgee Fields Incised; Complex I lacks this type. There was a noticeable lowering in the quality of the pottery made by the Historic Creek but this trait is to be expected, considering the fact that trade goods were entering the area. Sites of this cultural horizon were still common in the Chattahoochee Valley, 44 being represented. Many of the sites appeared large but their exact size was hard to determine because sherds, artifacts, and other surface indications were rare.

After the 1947 survey, intensive excavations were carried out at the Kolomoki Mounds north of Blakely, Georgia, only a short distance east of the Chattahoochee. This work was sponsored by the University of Georgia and the Department of State Parks in Georgia. William H. Sears, who directed the project, has analyzed his material and published his results on the four seasons of excavation in a series of three reports (Sears 1951, 1952, and 1953). To a certain extent, it is difficult to correlate the results of the 1947 survey in the Chattahoochee Valley with the cultural complexes defined by Sears, as the laboratory analysis of the 1947 survey data had already been completed and the pottery types defined before Sears undertook his project.

The Kolomoki Mounds are assigned by Sears to several distinct culture complexes, the oldest one being Weeden Island. This complex in turn was followed in time by a Kolomoki complex characterized by a variant of Swift Creek Complicated Stamped pottery that Sears has typed as Kolomoki Complicated Stamped (Sears 1951: 17). A trait of significance in the Kolomoki complex was the absence of check stamped pottery. All the sites found in 1947 by the writer that had on them Swift Creek Complicated Stamped sherds to the exclusion of check stamped types, were associated with a very crude plainware that was shared with the earlier sites related to the Deptford complex.

If Sears has correctly placed the Kolomoki cultural complex as following the Weeden Island I complex (defined by Willey and Woodbury as lacking check stamped pottery), a major problem is involved in the relationship of the Kolomoki complex to the Weeden Island II phase (characterized by the presence of check stamped pottery). According to Sears' interpretation, the Kolomoki complex would be contemporaneous with the Weeden Island II phase of the northwest coast of Florida. The 1947 survey established beyond reasonable doubt that sites characterized by check stamped pottery and other Weeden Island ceramics (refer to Ceramic Complex E in this report) existed in large numbers within a few miles of the Kolomoki site. If the Kolomoki complex and Weeden Island II phase were approximately contemporaneous, why weren't check stamped sherds found in significant numbers in the Kolomoki complex?

In 1945 Franklin and Barbara Fenenga made a survey along Bull Creek in the vicinity of Columbus, Georgia. A total of 15 village sites were located. The results of their survey have not been published but have been incorporated in a detailed manuscript. Rather than classifying them into the recognized types defined by other Southeastern archaeologists, they proposed four pottery types based on temper type. As a consequence, it is difficult to correlate their findings with those of the 1947 survey of the Chattahoochee Valley. Judging by the descriptions and illustrations in their manuscript, one gathers they found several single and multicomponent sites that differ little in their cultural content from those found in the 1947 survey and assigned to Ceramic Complexes B to I (Fenenga, Franklin and Barbara 1945).

In May 1940 a few excavations were made at the Coosa site by the supervisory staff of the Work Projects Administration Archaeological Project, sponsored by the Alabama Museum of Natural History. The Coosa was one of the most important sites of the Historic Creek Indians and was supposed to have been located on the Coosa River in Talladega County, Alabama. This site apparently was visited by the De Soto expedition in 1540 and was not abandoned until about 1832. Judging by the illustrations from site 1 Ta 1 (tentatively identified as Coosa), one concludes that there were many ceramic traits shared by cultures here and at the Abercrombie Village, site 1 Ru 10 (1 Ru 61). The major difference was a predominance of shell tempering at Coosa and a predominance of grit tempering at site 1 Ru 10 (1 Ru 61).

Ripley P. Bullen of the Florida Board of Forestry and Parks in 1948 made a survey of the Chattahoochee Valley from the Florida-Alabama border southward. For the most part, he found Fort Walton sites on which were a small amount of check stamped pottery, although he also encountered sites that ranged in time from Deptford to Historic Creek. He calls attention to the virtual absence of stoneware in the Fort Walton sites he tested (Ripley P. Bullen, *personal communication,* December 31, 1948).

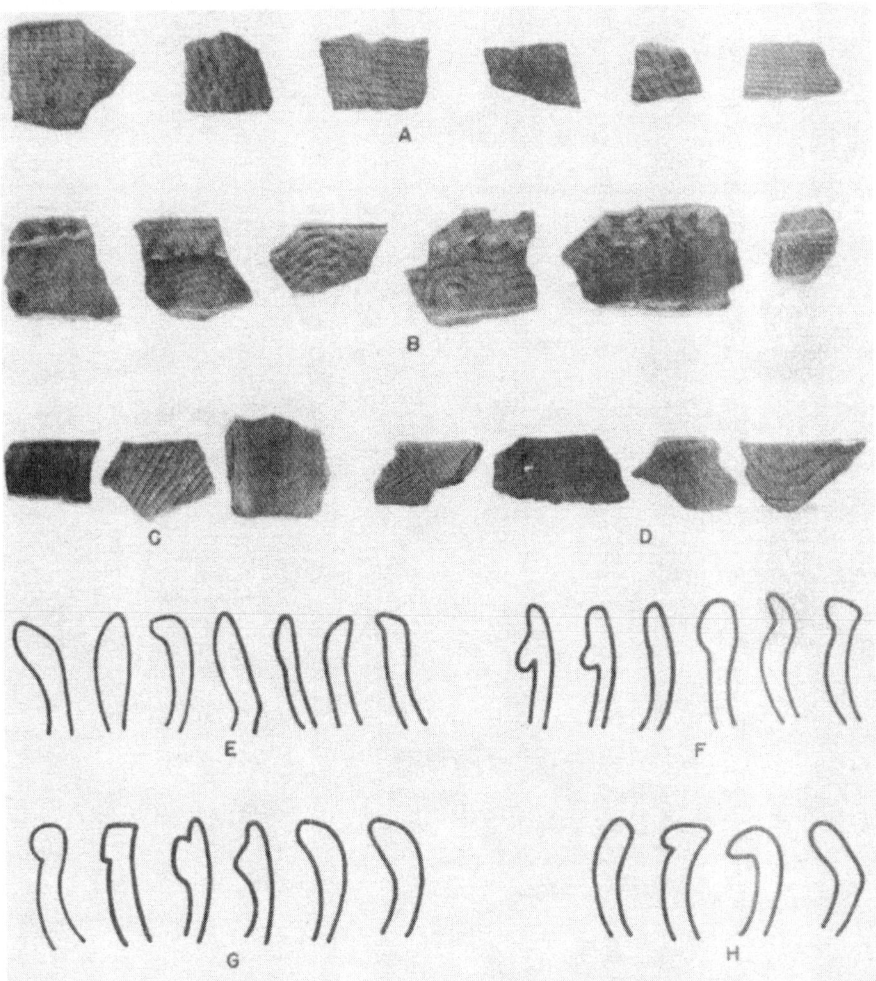

A- WILSON CHECK STAMPED SHERDS B - LAMAR COMPLICATED STAMPED SHERDS C- CHATTAHOOCHEE BRUSHED SHERDS D- COLUMBIA INCISED SHERDS E- WILSON CHECKED STAMPED RIM PROFILES F- LAMAR COMPLICATED STAMPED RIM PROFILES G- CHATTAHOOCHEE BRUSHED RIM PROFILES H- COLUMBIA INCISED RIM PROFILES

Figure 8 Miscellaneous Ceramic Types

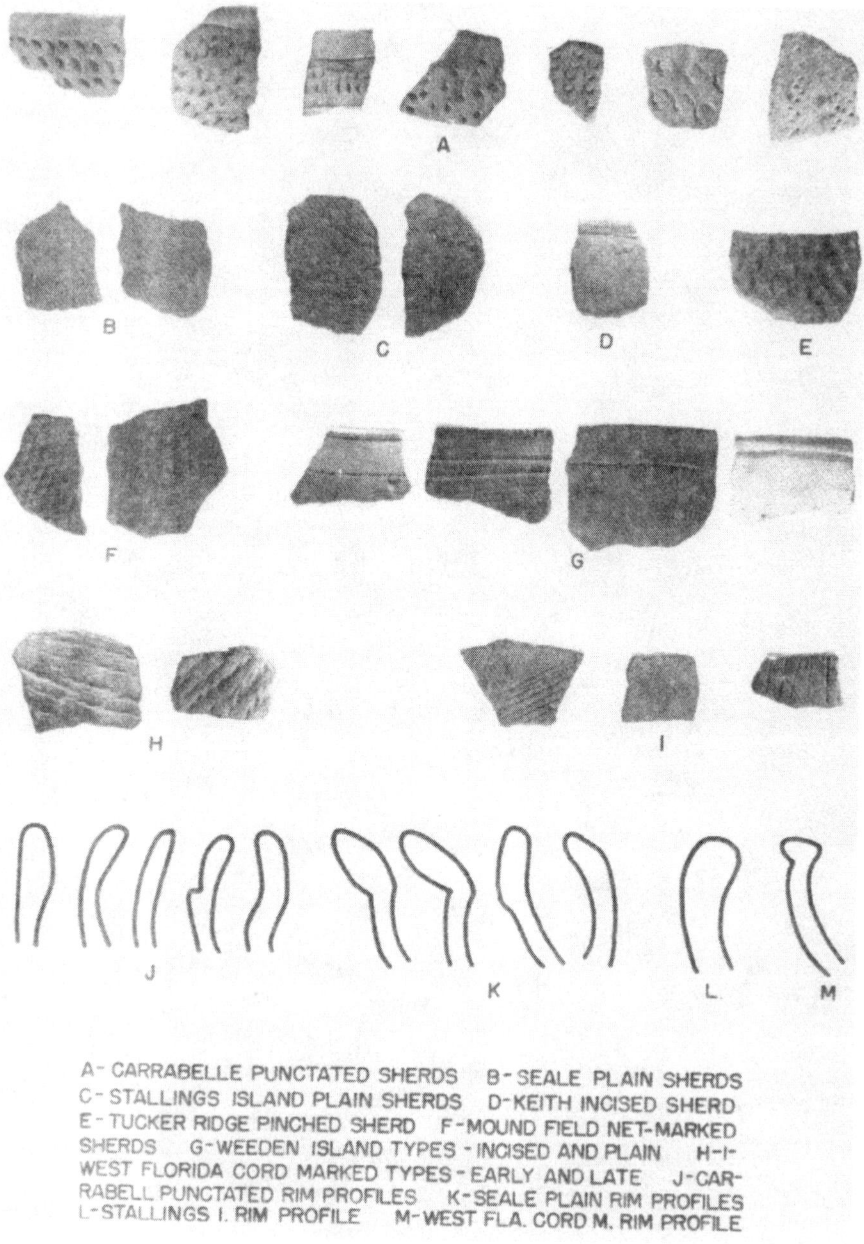

Figure 9 Miscellaneous Ceramic Types

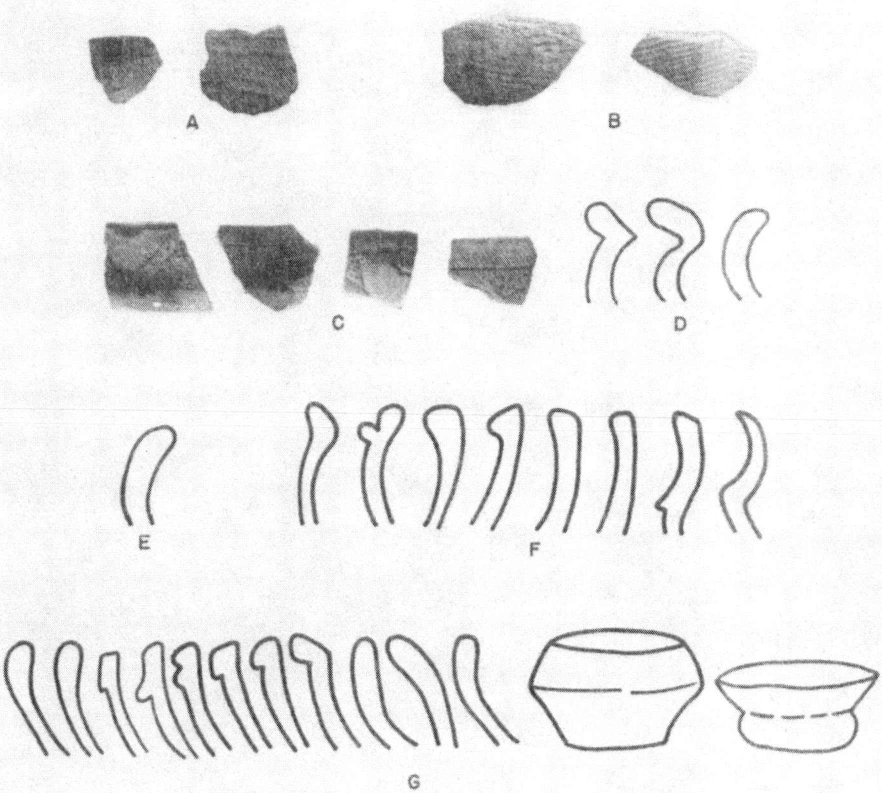

A-DEPTFORD BOLD CHECK STAMPED SHERDS B-DEPTFORD LINEAR CHECK STAMPED SHERDS C-FORT WALTON INCISED SHERDS D-DEPTFORD BOLD CHECK STAMPED RIM PROFILES E-DEPTFORD LINEAR CHECK STAMPED RIM PROFILE F-FORT WALTON INCISED RIM PROFILES G-COWETA MICACEOUS UTILITY RIM PROFILES AND VESSEL FORMS

Figure 10 Miscellaneous Ceramic Types

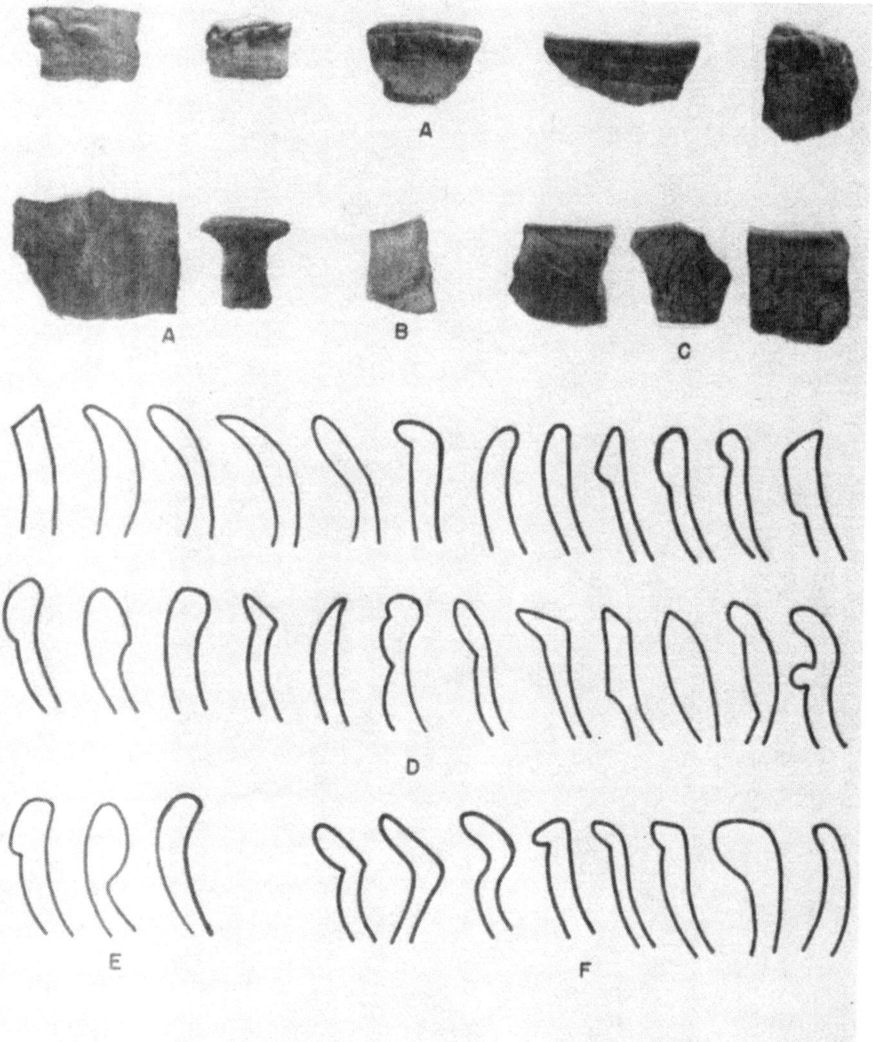

A-COLUMBIA UTILITY AND LAKE JACKSON PLAIN SHERDS B-KASITA RED FILMED SHERD C-OCMULGEE FIELDS INCISED SHERDS D-COLUMBIA UTILITY AND LAKE JACKSON PLAIN RIM PROFILES E-KASITA RED FILMED RIM PROFILES F-OCMULGEE FIELDS INCISED RIM PROFILES

Figure 11 Miscellaneous Ceramic Types

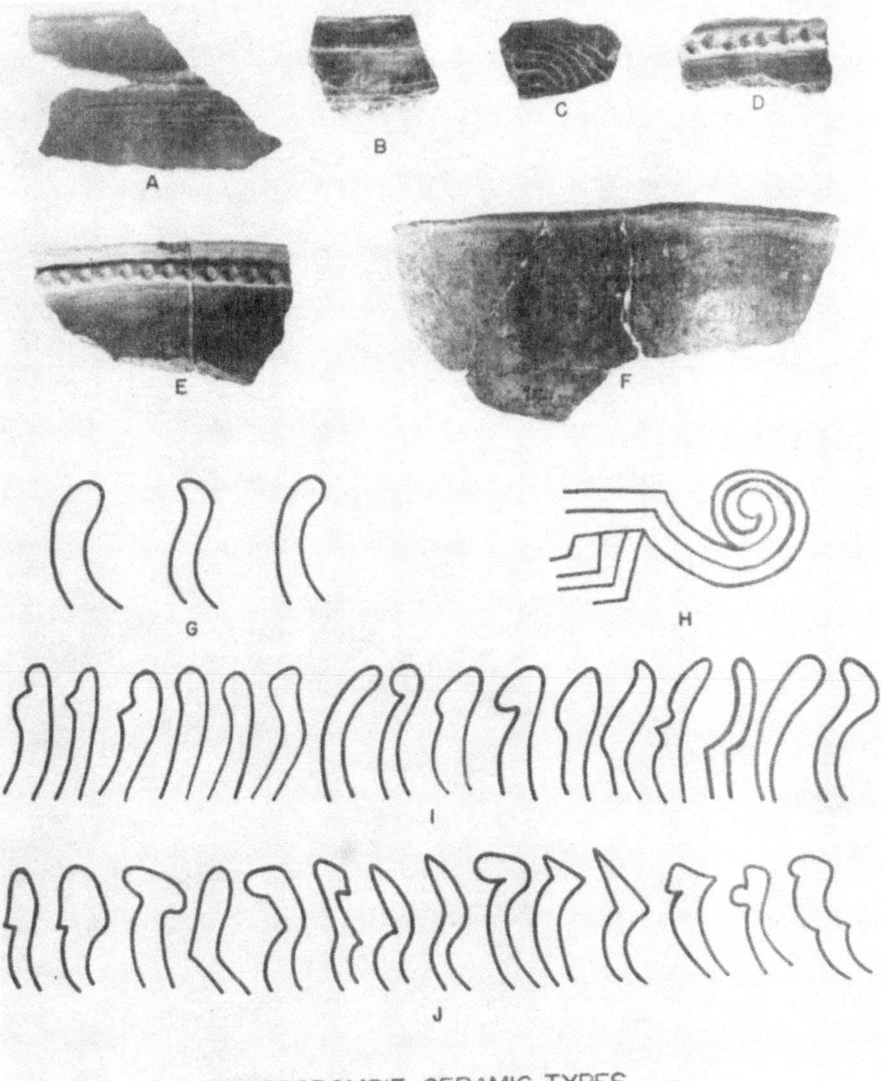

ABERCROMBIE CERAMIC TYPES
A-C-ABERCROMBIE INCISED, D-F-ABERCROMBIE PLAIN
G-ABERCROMBIE INCISED RIM PROFILES, H-ABER-CROMBIE INCISED DESIGN, I,J-ABERCROMBIE PLAIN RIM PROFILES.

Figure 12 Abercrombie Ceramic Types

Figure 13 Projectile Point Types

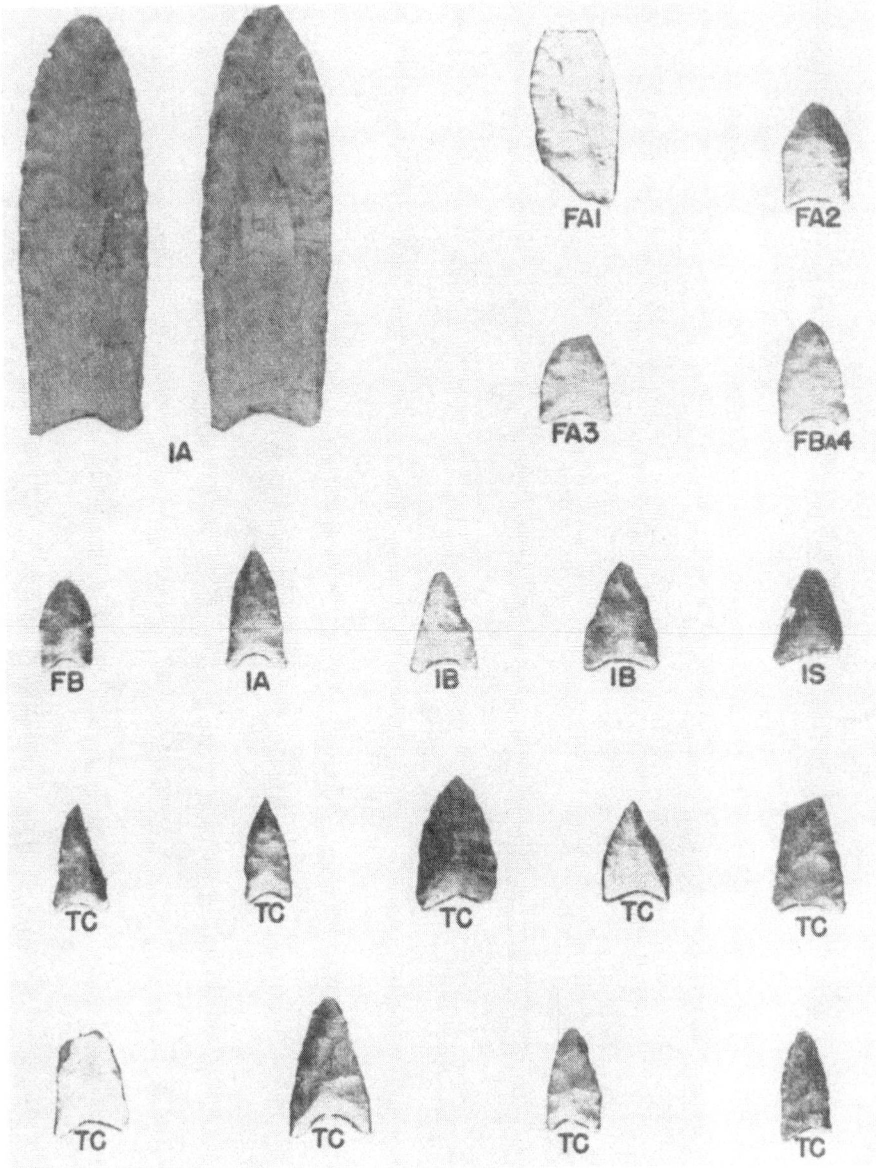

Figure 14 Projectile Point Types

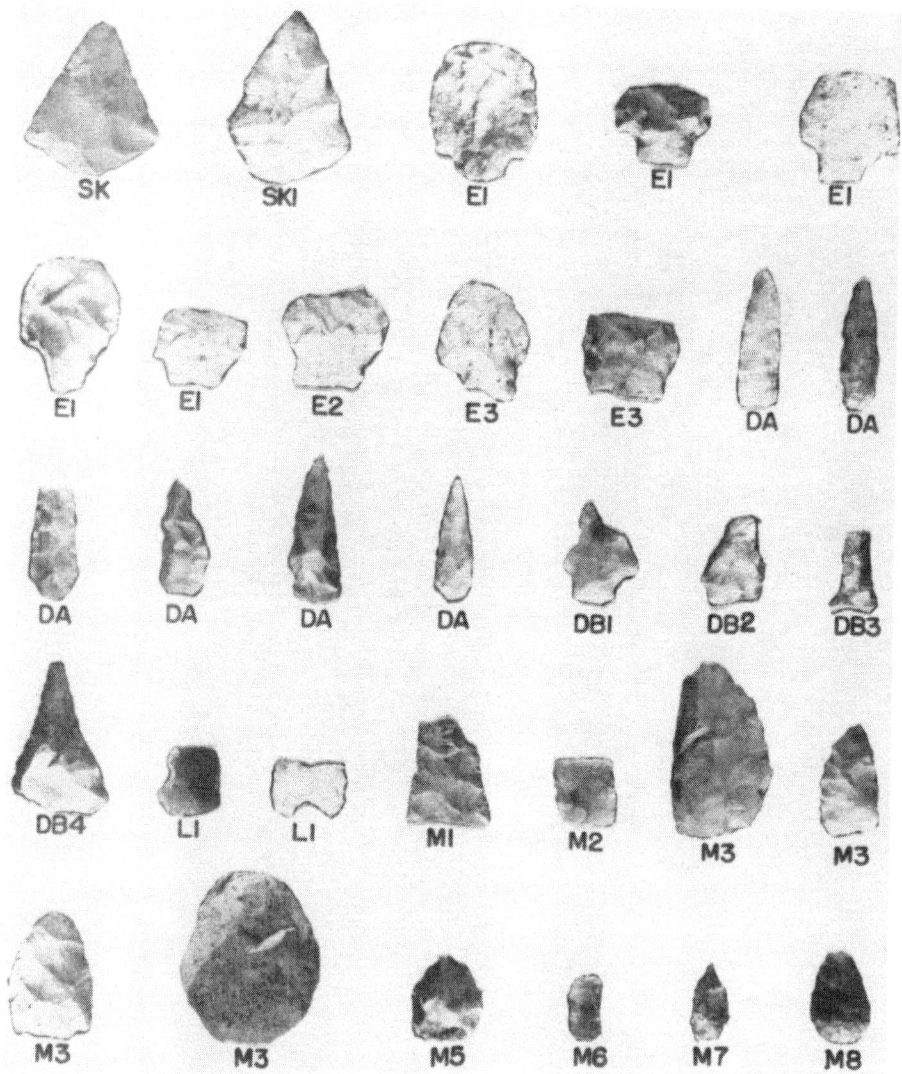

Figure 15 Flint Tool Types

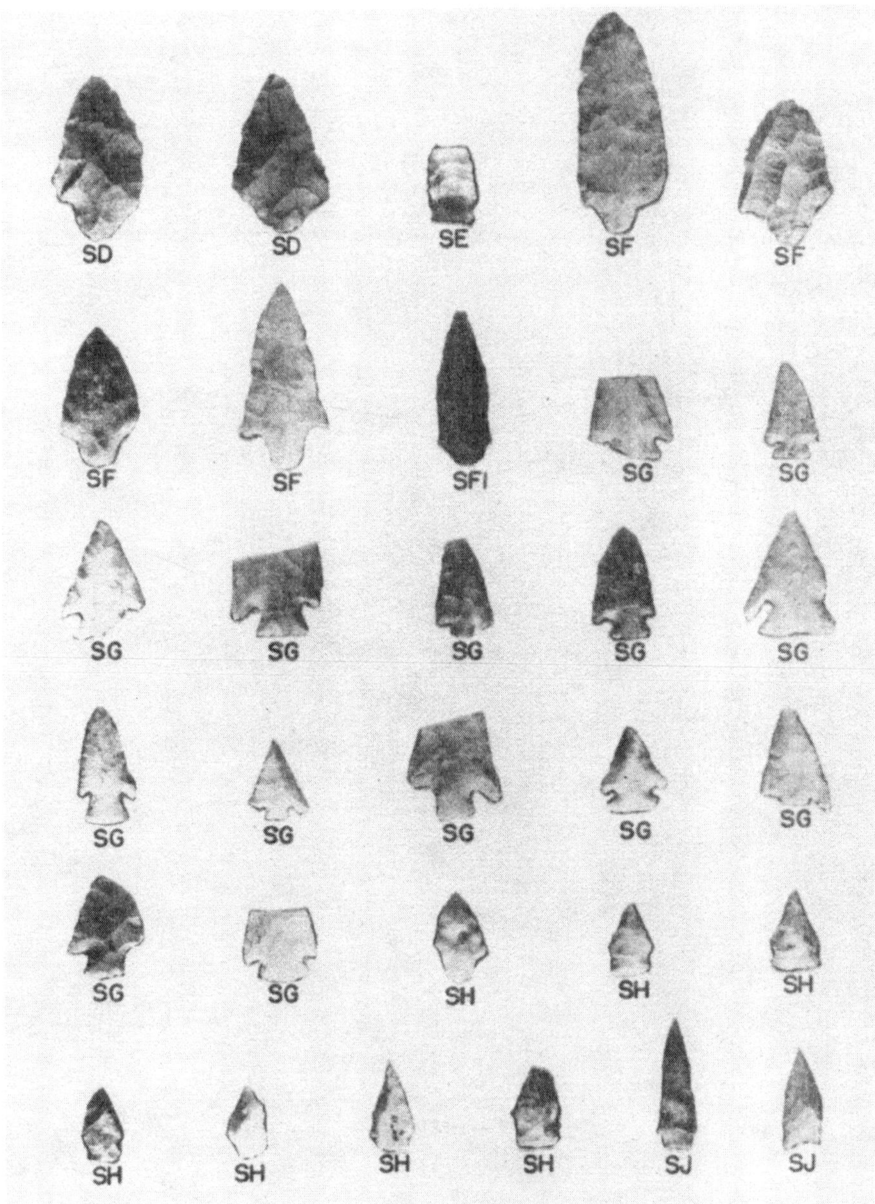

Figure 16 Projectile Point Types

Part II

ARCHAEOLOGICAL INVESTIGATIONS IN THE WALTER F. GEORGE BASIN

Edward B. Kurjack

CHAPTER 4

Excavations

1 BR 15, THE SHORTER SITE

The Shorter Site, 1 Br 15, consisting of a high but eroded mound and adjacent village area, is located just north of Eufaula, Alabama, less than a half mile northwest of the Chattahoochee River. Here, on the second terrace of the river, three sites lie parallel to the Chattahoochee. Site 1 Br 15 is centermost, bounded on both the northeast and southwest by deep gullies cut by two branches of the river. These gullies and the heavy woods growing along their sides separate the site from 1 Br 35 on the southwest and 1 Br 37 on the northeast. Prior to acquisition by the U.S. Corps of Engineers, the area had been extensively farmed. Cotton grew there during the first excavations.

Excavation of the site began in June 1960 with a crew of University of Alabama students. Two units were excavated: a test trench in the village area and a major exploration of the mound, 1 Br 15X1. The excavation of the mound was continued in the fall of 1961.

The site was rich in surface material and a large collection, consisting mostly of pottery, was gathered. The collection revealed the presence of five components:

(1) A Historic Creek or Ocmulgee Fields component, represented by brushed pottery and a few European trade objects.

(2) A Fort Walton-Lamar component represented by the types Fort Walton Incised, Lake Jackson Plain, Lamar Complicated Stamped, and other common pottery.

(3) An early Swift Creek component represented by the early varieties of the type Swift Creek Complicated Stamped and related Santa Rosa-Swift Creek wares.

(4) A Cartersville component represented primarily by check stamped sherds and by several related types.

(5) A small fiber tempered late Archaic component represented by large, stemmed projectile points and fiber tempered sherds.

A frequency distribution of the various ceramic types in the surface collection appears in Table 22.

The Lamar Plain category, ideally consisting of body sherds from Lake Jackson Plain vessels or undecorated portions of Lamar Complicated Stamped or Fort Walton Incised vessels, here probably includes undecorated fragments of Ocmulgee Fields vessels as well. The shell tempered ware was listed with the Fort Walton-Lamar types because of the lack of Walnut Roughened sherds in the collection. The Micaceous Plainware category includes plain sherds with ware characteristics similar to Cartersville Check Stamped. Other artifacts are classified below by components.

Ocmulgee Fields Component: One English claypipe stem fragment (4/64 inch bore diameter); one gun flint; one glass wine bottle fragment; one ball of lead shot.

Lamar Component: Three long, thin, narrow, triangular projectile points (related to the Mississippian Triangular types); five fragments of long, thin, and relatively wide, triangular points, all too large to have been used as arrowheads; one greenstone celt fragment.

Archaic Component: Three large, stemmed projectile points; one ovate uniface scraper.

Miscellaneous: One reel-shaped plate gorget fragment; two painted chinaware fragments; one fragment of worked hematite.

Wesley R. Hurt made a surface collection at 1 Br 15 during his 1947 survey of the area. He made note of the multi-component nature of the site, classing the mound (which he termed 1 Br 2) as a Fort Walton-Lamar manifestation. Hurt considered the check stamped ware at the site to be an early variety of Wakulla Check Stamped. He thought the Early Swift Creek material and the check stamped pottery were part of a single Weeden Island component.

THE VILLAGE AREA

The field exploration of the village area at 1 Br 15 was accomplished by a single, long trench, dug during the summer of 1959 by students under the direction of David L. DeJarnette. The trench, measuring 5 feet in width and 80 feet in length, was located approximately 250 feet northeast of the mound. Enlargement of the trench at the end nearest the mound was necessitated by the detection of two large refuse pits in the upper half of the trench.

The site was composed of a white sand extending to a depth of 4 to 6 feet, and overlying sandy, yellow clay was present.

Cultural material was found to a depth of three feet in this part of the village area, and excavation proceeded to this depth by arbitrary 4-inch levels.

Figure 17 An Aerial View of the Shorter Site, 1 Br 15

TABLE 15

Pottery Found on the Surface of 1 Br 15

		Number	Percent
Ocmulgee Fields Types			
Chattahoochee Brushed		43	3.258
Kasita Red Filmed		2	.152
Ocmulgee Fields Incised		1	.076
	Total	46	3.486
Fort Walton-Lamar Types			
Fort Walton Incised		15	1.136
Lake Jackson Plain		13	.985
Plain Shell Tempered		11	.833
Lamar Complicated Stamped		5	.379
Lamar Plain		187	14.167
	Total	231	17.500
Early Swift Creek Types			
Swift Creek Complicated Stamped (Early Variety)		13	.985
Crooked River Complicated Stamped		13	.985
West Florida Cord Marked (Early Variety)		8	.606
	Total	34	2.576
Cartersville Types			
Cartersville Check Stamped		462	35.000
Deptford Simple Stamped		12	.909
Deptford Linear Check Stamped		8	.606
Micaceous Plainware		459	34.773
Dunlap Fabric Marked		29	2.197
	Total	970	73.485
Miscellaneous			
Stallings Plain		3	.227
Miscellaneous Zone Punctated Ware		3	.227
Eroded Complicated Stamped		14	1.061
Residual Plain		19	1.438
	Total	39	2.953
	SITE TOTAL	1320	100%

Figure 18 The Test Trench in the Village Area at 1 Br 15

Figure 19 Excavation of Refuse Pits, Features 15 and 16, in the 1 Br 15 Test Trench

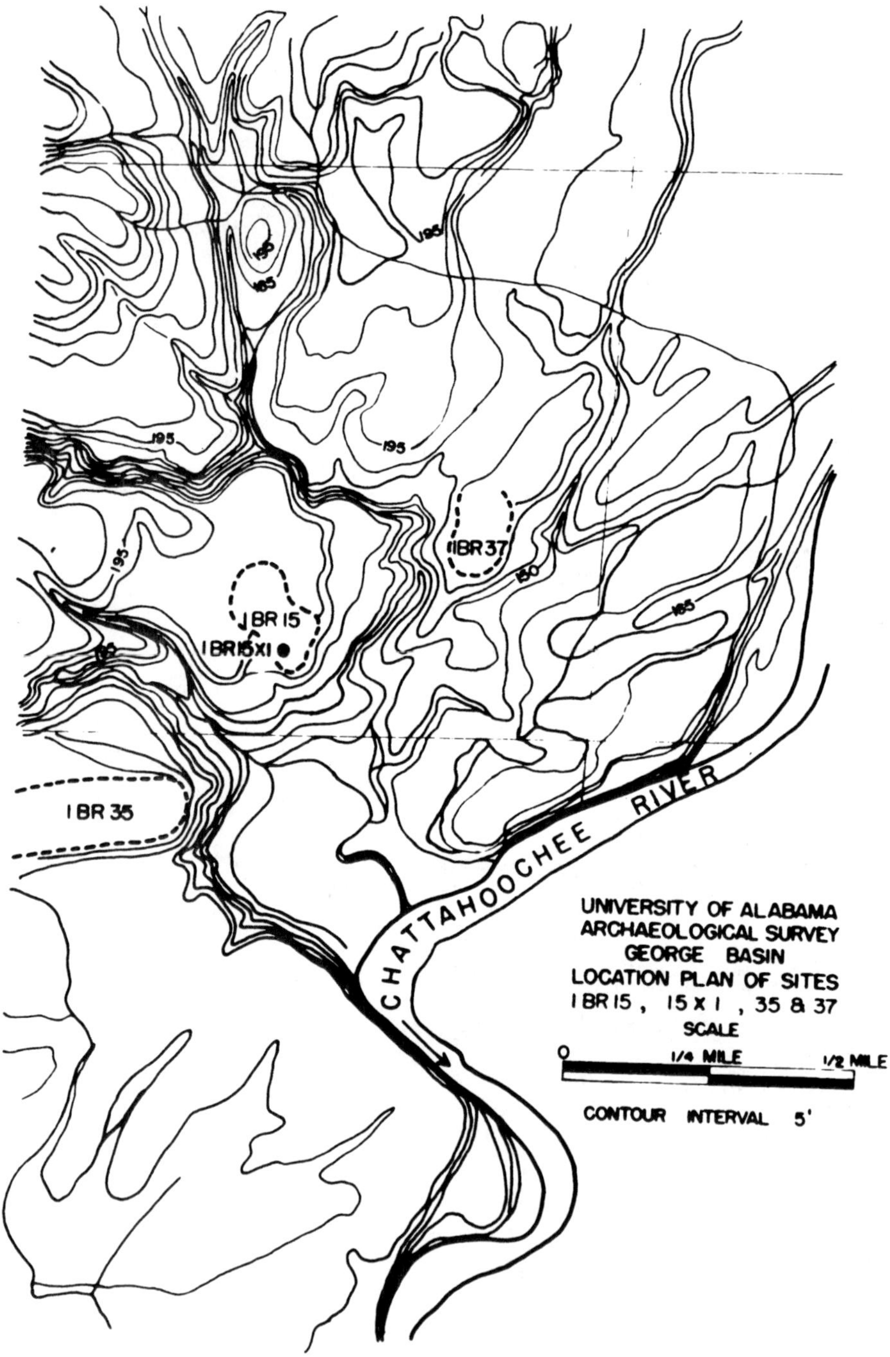

Figure 20 University of Alabama Archaeological Survey
George Basin; Location Plan of Sites
1 Br 15, 15X1, 35, and 37

FEATURES

Five types of features were uncovered in the test trench in the village area. Best defined of these were five large refuse pits, resembling Feature 15 (shown in the lower left of Figure 19). All of the large refuse pits contained some Fort Walton-Lamar or Ocmulgee Fields pottery, indicating that they were dug by the more recent, Mississippian occupants of the site. A sixth pit was similar in form and content to the small, shallow pits filled with charred corncobs found at Ocmulgee Old Fields sites such as 1 Br 35. Eight small concentrations of pottery fragments, quartzite pebbles, charcoal and flint debris were noted. These small "miscellaneous features" seldom showed any indication of pit outlines. In addition, two well-defined animal burrows–Features 7 and 9–were located.

Pottery was the only type of artifact found in the features, although flint debris, quartzite pebbles, and charcoal were invariably present. A description of the pottery from each feature is given in Table 16.

POTTERY

The large variety of ceramic types present at 1 Br 15 was indicated by the surface collection. Large collections of sherds were taken from both the test trench in the village area of 1 Br 15 and the excavations in Mound 1 Br 15X1. Each of the pottery types or categories found in the various areas of the site is described briefly in the following paragraphs.

Ocmulgee Fields Types

Chattahoochee Brushed (Bullen 1950). Several small sherds of this coarse grit tempered, brushed or cob marked ware were present at the site. Although constituting most of the Historic Creek pottery from the site, this type is almost completely absent in the pottery collections from Mound 1 Br 15X1.

Ocmulgee Fields Incised (Jennings and Fairbanks 1939). A single, weakly incised casuela rim with an extruded lip was found on the surface.

Kasita Red Filmed (Anon. 1948b). A few sherds, usually cream-colored and painted with a deep red hematite film, were present on the surface and in the village area.

Ocmulgee Fields Plainware. Plain body sherds from Ocmulgee Fields vessels were not separated from the undecorated body sherds of the Fort Walton-Lamar types.

Fort Walton-Lamar Types

Lake Jackson Plain (Willey 1949). A few coarse grit tempered rim sherds were collected at the site. A variety of strap handles, loop handles, nodes, and lugs were present, as well as the common, pinched or indented fillet. Rims varied in form from straight to thickened, flaring or bent at sharp angles to the shoulder.

TABLE 16
Pottery from the Features in the 1 Br 15 Test Trench

	Large Refuse Pits					Cob Hole	Miscellaneous Features					Animal Burrows	
Feature Number	5	10	12	15	16	1	2	3	4	8	13	7	9
TYPE													
Chattahoochee Brushed			2		4							1	
Kasita Red Filmed													
Lamar Complicated Stamped								2					
Lake Jackson Plain			1					1	1				
Plain Shell Tempered			3	10									
Fort Walton Incised									1				
Lamar Plain	3	12	60	24	20	8		5	5		6		5
Early Swift Creek Complicated Stamped		1	1									1	1
West Florida Cord Marked	3		8	7	8			1			1		1
Cartersville Check Stamped	7	1	15	10	5	2		6	4		6	9	
Deptford Simple Stamped									1		3		2
Deptford Bold Check Stamped	1		19										
Micaceous Plainware	14	5			6	1		4	2		1		3
Stallings Plain	1		2										
Residual Plain								1		2			
TOTAL	29	19	111	51	43	11		20	14	2	17	11	12

Fort Walton Incised (Willey 1949). These sherds were usually small body fragments decorated with punctations zoned by curvilinear, incised lines.

Lamar Complicated Stamped (Jennings and Fairbanks 1939). A small number of coarse grit tempered, complicated stamped body sherds with weak paddle impressions were found; no rim sherds were present. The conspicuous absence of incised casuela rim sherds from the collection indicates that vessels were of other forms.

Pinellas Incised (Willey 1949). A single sherd with a notched exterior lip and incised lines parallel to the rim was taken from the mound.

Plain Shell Tempered Ware. Shell tempered sherds occurred frequently but all were undecorated. The lack of a single brushed, shell tempered sherd indicates that the shell tempered ware is related to the Fort Walton-Lamar component.

Lamar Plain. Grit tempered plainware from the site was separated into two categories: hard, coarse grit tempered body sherds from Lake Jackson Plain and Fort Walton Incised vessels, and fine sand and mica tempered sherds similar in ware characteristics to the Cartersville Check Stamped ware. The Lamar Plain category thus includes Ocmulgee Fields Plainware.

Early Swift Creek Types

Swift Creek Complicated Stamped (Early Variety) (Willey 1949). Several early, curvilinear, complicated stamped sherds were present in the various collections from the site. On a few sherds, check stamped areas were integrated with curvilinear elements of the design. Most of the rims examined were of the typical Early Swift Creek notched variety. Tetrapodal supports were also present. The ware characteristics graded from the fine sand and mica paste characteristic of the Cartersville ware at the site, to a slightly harder, finer, more compact material.

Crooked River Complicated Stamped (Willey 1949). A few rectilinear, complicated stamped body sherds with chevron designs were found.

West Florida Cord Marked (Willey 1949). Two types of cord marked ware appeared at the site. The most common variety was characterized by the closely spaced, parallel impressions of fine cords. The cord impressed on the second variety was slightly larger, and the impressions were randomly applied, leaving large areas of unmarked surface. Rim sherds from cord marked vessels were few in number; those that were examined possessed a slight rim fold.

Cartersville Types

Cartersville Check Stamped (Caldwell 1958). Sherds of this ware were the most numerous at the site. The type is characterized by fine check stamping, impressed on conoidal or truncated vessels. The lands forming the check design were, on the average, 1.5 to 3 mm. apart. Often the lands in one direction were higher than those in another. Three styles of vessel bases were present: (1) a truncated base with medium to large tetrapodal supports; (2) a thick, squared, truncated base; and (3) a rounded, conoidal base. Three rim styles were used: (1) a simple rim above a 3- or 4-inch, unstamped zone; (2) a simple rim with check stamping extending up to the lip; and (3) a slightly folded rim with check stamping extending up to the fold. Examples of this type are shown in Figure 27.

Cartersville Check Stamped sherds at 1 Br 15 were tempered with fine sand and mica. Although the ware was not hard and surfaces were often eroded, the surfaces were well smoothed on the uneroded specimens. Color ranged from buff-red to the more common black.

Cartersville Simple Stamped (Caldwell 1958). A large group of simple stamped body sherds was collected. Most of these appeared to have been marked with the impressions of a simple, grooved paddle. The few rim sherds noted showed slight rim folds. Tetrapodal supports were present.

Dunlap Fabric Marked (Sears and Griffin 1950b). A few fine sand and mica tempered sherds were decorated by stamping with a cord-wrapped dowel, or perhaps impressed with a plain, plaited fabric. Although no rims were observed, tetrapodal supports were present.

Deptford Bold Check Stamped (Willey 1949; Griffin and Sears 1950). A small number of bold check stamped sherds were present. Lands appearing on this type averaged approximately 5 mm. or more apart. The rims observed on this type were straight and simple, without a folded lip.

Micaceous Plainware. Many body sherds with ware characteristics similar to those of Cartersville Check Stamped sherds were found at 1 Br 15. A few straight, simple rims were present. Sherds in this category included, for the most part, the undecorated portions of Cartersville vessels, eroded Cartersville sherds, and fragments of a few plain vessels.

Fiber Tempered Types

Stallings Plain (Sears and Griffin 1950a). A few fiber tempered body sherds were collected. No rims or bases were found.

Stallings Punctate (Sears and Griffin 1950a). Two fiber tempered, linear-punctate sherds were found in the tertiary mound fill.

Pottery Depth Distribution in the Test Trench

The depth distribution of pottery types in the village test trench is reproduced in Figure 21. The diagram shows little significant change in relative frequency of type with depth; this is probably due to undetected mixing at the site.

EXCAVATIONS IN THE MOUND, 1 Br 15X1

The excavation of the mound 1 Br 15X1 proceeded in four steps. After the mound was cleared, a small test trench was excavated at its center. (The loose sand encountered at the top of the mound made this work hazardous; before the submound of the site was reached, the trench was almost completely filled by the caving sides. Therefore, because the site was scheduled to be completely flooded, earth-moving equipment was utilized.) A small bulldozer was employed to cut a swath through the site. After the profile cut by the machine was cleaned and examined, excavations were made in various places in order to sample the various zones of the site. The site was revisited in the fall of 1961, when the top zone of the site was removed on one side of the 1960 excavations in order to search for structures.

Profiles

A contour map of the mound shown in Figure 22 indicated the possible presence of a flat-topped, rectangular mound. Excavation of the site, however, provided the view of the profiles illustrated with the map and in the photograph in Figure 25. Four zones, described below, were present.

The Tertiary Mound. This was the final stage of mound-building at the site. The surface was heavily eroded, and little remained of the once flat top. The zone was composed of white sand and gravel, with clay lenses and water seepage lines present. Pottery and artifacts were abundant within this zone.

The Secondary Mound. A flat, clay cap was found over the primary mound and under the tertiary. A large area of the secondary mound floor was uncovered on the north side of the site during the second season's operations. The sand of the tertiary mound was easily removed from the hard clay, secondary mound surface. This surface was carefully examined for evidence of structural remains, but no post molds were detected. The hard clay at the top of the secondary mound included more sand lenses in its deeper portions. A large pit cut through the secondary mound during the second season's work yielded only a handful of pottery fragments; cultural material was very scarce in the secondary mound fill.

The Primary Mound. The initial stage of construction was the building of a relatively flat-topped mound of sand. A thin layer of charcoal marked the top of the primary mound; in one area this layer was 5 inches thick. The sides of the primary mound were straight, as shown in the profile drawings and Figure 26. The primary mound was constructed of fine, gray-white sand, identical to the

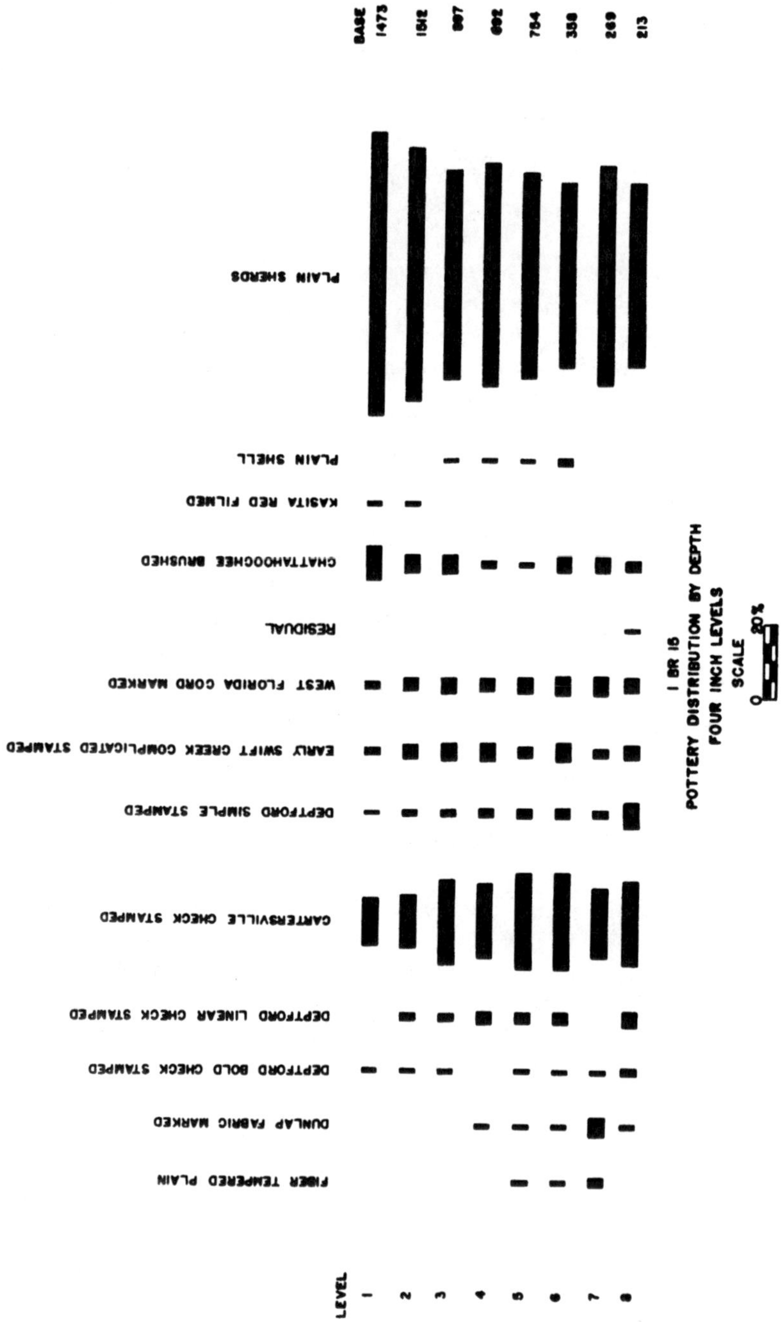

Figure 21 Pottery Distribution by Depth, 4-Inch Levels

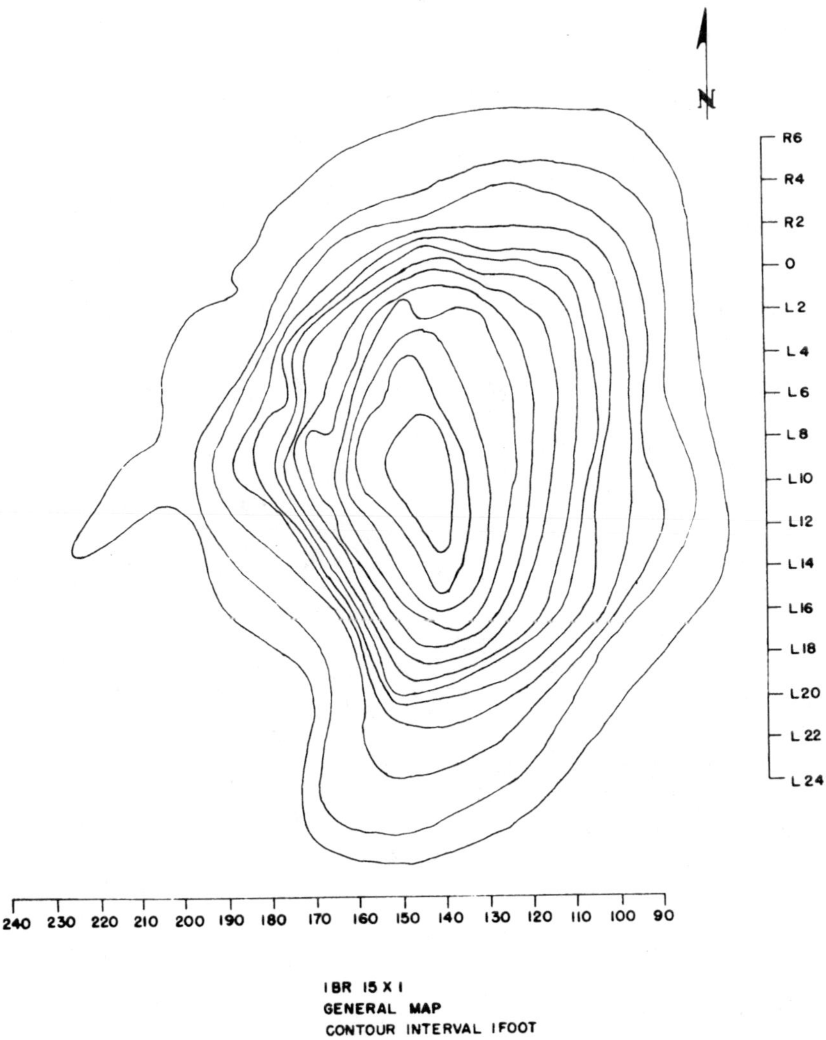

Figure 22 1 Br 15X1 General Map

Figure 23 Operation of the Bulldozer on the Shorter Mound, 1 Br 15X1

sand on the surface of the village surrounding the mound. Sherds were frequently found within this layer.

The Pre-Mound. The old humus line, marked in most areas by a thin layer of charcoal, was clearly visible below the primary mound. This line neatly separated the pre-mound zone from the primary mound. The pre-mound, composed of gray-white sand, yielded few sherds below the depth of one foot. A fairly large submound ceramic sample was collected.

Stratigraphy

The pottery types found within the mound are described with the wares from the village area. The occurrence of pottery from the various areas of the mound is shown in Tables 17 and 18. Two principal collections were made: sherds taken from the first test trench excavated through the center of the mound, and pottery collected from the various zones after the profiles were cut through the mound. Scarcity of material in the secondary mound is reflected in the absence of sherds in the test trench at depths between 5 and 9 feet.

A single fact concerning the depth distribution of the pottery quickly became apparent: No Fort Walton-Lamar types were present below the tertiary mound. The super-position of the Fort Walton-Lamar component is thus easily demonstrated at the mound, while the virtual absence of Ocmulgee Fields types within the mound effectively separates the Historic Creek component at the site from all previous cultural remains.

Features

Several features were defined during the excavation of the mound. While isolated post molds, animal burrows, and tree root systems were plentiful, the features of special interest are described below:

Feature 4. The location of this refuse pit is shown on the profile drawing of the mound. The pit, exposed by the profile cut, was approximately 3 feet in diameter and 14 inches deep. The feature contained many sherds from a single Cartersville Check Stamped vessel. (One other sherd was classified as Lamar Plain.) In addition, a small, triangular projectile point (4.2 mm. long and 2.2 mm. wide) with serrated edges and a deeply concave base was taken from the feature.

Features 6 and 7. Two adjacent, relatively shallow pits were found under the south profile in the pre-mound zone. These pits contained a projectile point 5.5 mm. long and 2.0 mm. wide, identical in appearance to the one found in Feature 4. Four Cartersville Check Stamped sherds were also present.

Feature 16. This feature was a large refuse pit found under the south profile of the mound. The pit, detected in the pre-mound zone, measured approximately 3 feet in diameter and a foot in depth. It contained two broken projectile points

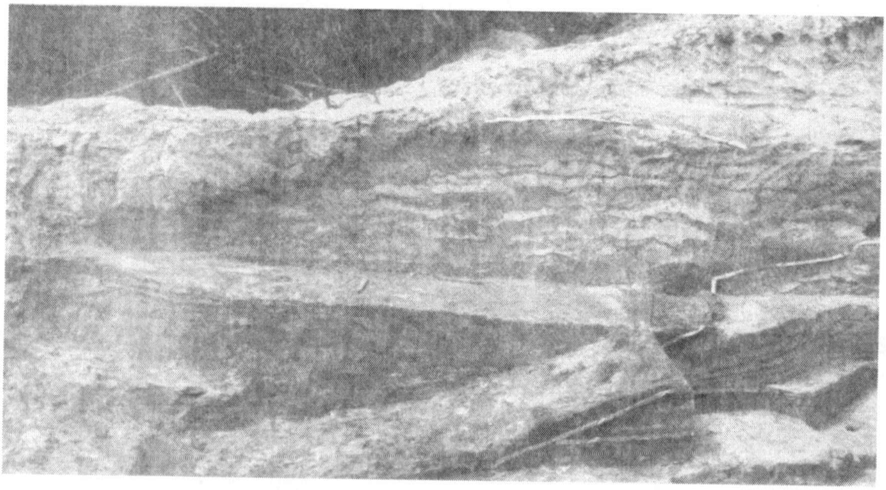

Figure 24 The West Side of the North Profile, 1 Br 15X1

Figure 25 A View of the North Profile of the Shorter Mound, 1 Br 15X1

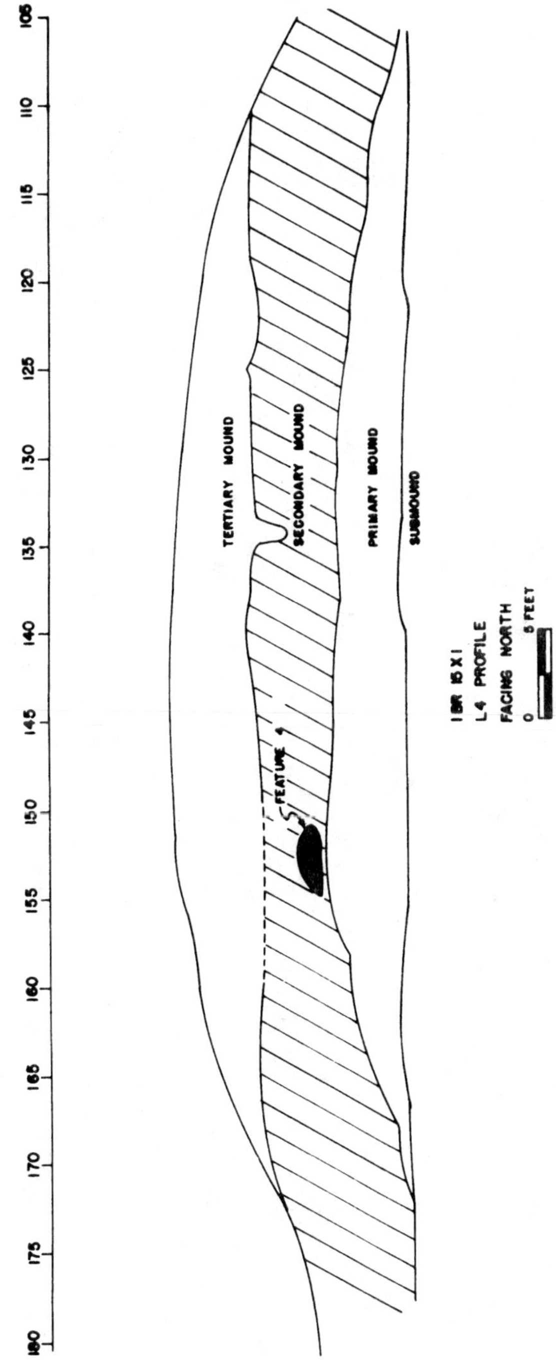

Figure 26 1 Br 15X1 – L4 Profile Facing North

TABLE 17

Pottery from the Test Trench on Mound 1 Br 15X1

TYPE	Depth in Feet							
	0-1	1-2	2-3	3-4	4-5	9-10	10-11	11-12
Lake Jackson Plain	3	5			2			
Fort Walton Incised	5	2	1		5			
Plain Shell Tempered		1			1			
Lamar Plain	34	17	4		24			
Swift Creek Complicated Stamped (Early Variety)	12	7	1					
Cartersville Check Stamped	26	42	14	7	16	17	30	3
Cartersville Simple Stamped	6	11			1			
Deptford Bold Check Stamped	1	4	1		1			
Dunlap Fabric Marked	1	1	2				1	1
Micaceous Plainware	46	41	13	5	13	2	13	2
Stallings Plain			1				1	
Residual		5	1		1			
Total	134	136	38	12	64	19	45	6

TABLE 18

Pottery from the Mound by Zones, 1 Br 15X1

	Tertiary Mound	Secondary Mound	Primary Mound	Pre-mound
Lake Jackson Plain	6			
Lamar Complicated Stamped	7			
Fort Walton Incised	13			
Pinellas Incised	1			
Plain Shell Tempered	5			
Lamar Plain	175	1		
Crooked River Complicated Stamped	1	1		
Swift Creek Complicated Stamped (Early Variety)	33			
West Florida Cord Marked	5			
Cartersville Check Stamped	373	35	117	550
Cartersville Simple Stamped	46			
Deptford Bold Check Stamped				1
Dunlap Fabric Marked	8		2	29
Micaceous Plainware	223	8	46	290
Miscellaneous Zone Punctated Ware				
Stallings Plain	6			
Stallings Punctated	2			
Residual	3			
TOTAL	907	45	165	870

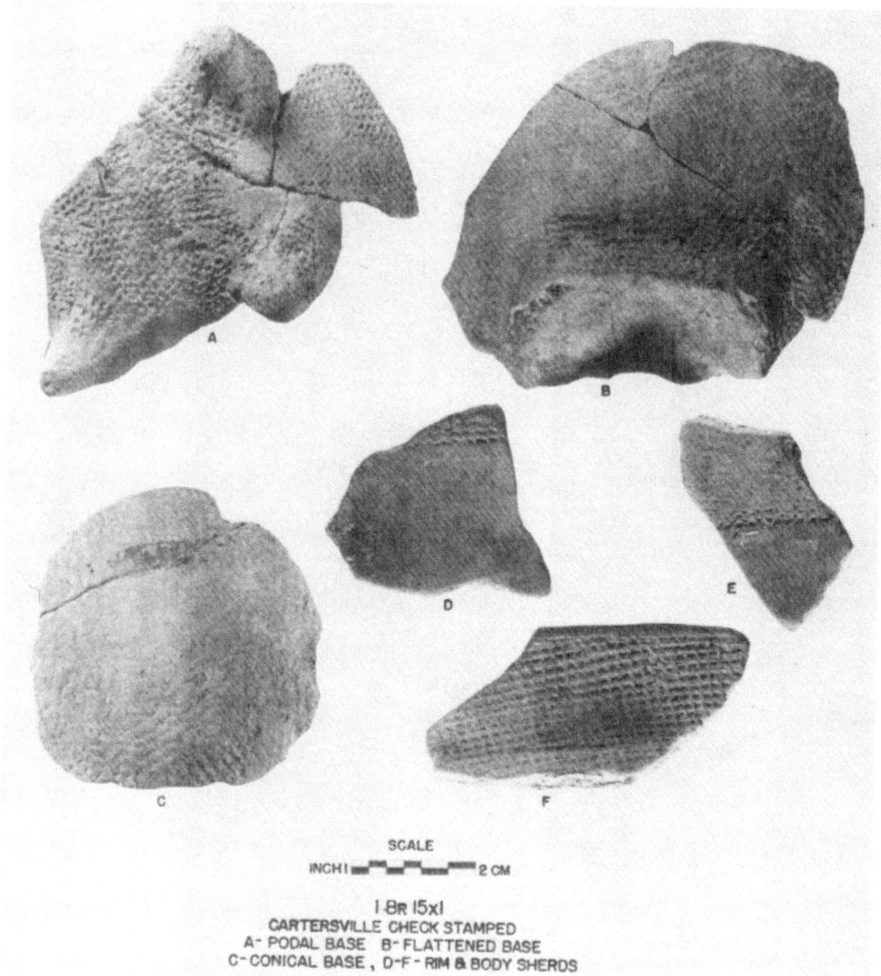

Figure 27 Cartersville Check Stamped, 1 Br 15X1

Figure 28 Uncovering the Top of the Secondary Mound During the Second Season's Work at 1 Br 15X1

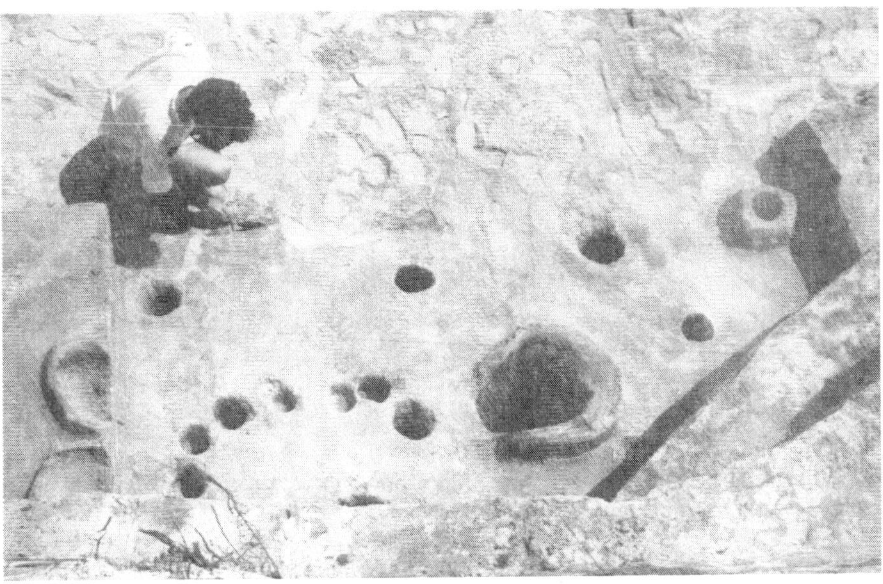

Figure 29 Feature 13, 1 Br 15X1. Post Molds in the Pre-Mound Zone near the Center of the Shorter Mound

of the type found in Features 4, 6, and 7; a fragment of limonite with a high metallic luster; and large sherds from a single Cartersville Check Stamped vessel.

Feature 10. This feature, probably a fire hearth, was a shallow pit 19 inches in diameter filled with quartzite pebbles. Six Cartersville Check Stamped sherds and two fine sand and mica tempered body sherds were also present.

Feature 13. This feature consisted of a group of post molds detected in the pre-mound zone near the center of the mound. The pattern of the posts, shown in Figure 29, could not be completed.

1 Br 15: NON-CERAMIC ABORIGINAL ARTIFACTS

In addition to the artifacts already mentioned from 1 Br 15 are three fragmentary projectile points, one lanceolate basal section, one distal end, and one bifacially chipped knife.

1 Br 15: MISCELLANEOUS

One iron spike of the wrought variety was found at the site. This find occurred in the plow zone.

1 Br 15X1: NON-CERAMIC ABORIGINAL ARTIFACTS

Five stone artifacts—four fragmentary projectile points and a fragment of a reel-shaped, hematite gorget—were collected at 1 Br 15X1. One of the projectile point fragments appeared to be a barbed-shoulder type; the others were distal ends. All four appeared to be Archaic.

SUMMARY: 1 Br 15

Five components were found at 1 Br 15, the earliest of which was represented by a few steatite sherds, stemmed projectile points, and fiber tempered pottery. No large component of this complex has yet been investigated, although small amounts of cultural material from the complex were present at many of the Chattahoochee Valley sites. The decoration on a few of the fiber tempered sherds suggests a relationship with the Stallings Island complex of the north Georgia coast.

Ceramic typology indicates that the Cartersville component was the largest at the site. High frequency of check stamped and simple stamped wares and small amounts of fabric marked pottery are evidence that the component dates from the latter part of the Cartersville Period. The similarity in ware characteristics of the Cartersville and the Swift Creek sherds at the site suggests a close relationship between the two components.

Little evidence concerning economic activity during the Cartersville Period was detected. The formative status of Deptford cultures has been hypothesized by Morse (1961). Caldwell (1958) notes a decline in the importance of the storage pits, which he considers evidence of an acorn-collecting economy during the Cartersville Period.

Remains of the Fort Walton-Lamar occupation are similar to those of the last occupation of the Rood's Mound (Caldwell 1955b). The present interpretation is that the three mound stages of 1 Br 15X1 were built during the Fort Walton-Lamar occupation.

The Ocmulgee Fields component at 1 Br 15 is related to the major component at nearby 1 Br 35. Dating from the middle of the eighteenth century, the component represents a Historic Creek occupation.

1 BR 35, THE JACKSON SITE

Site 1 Br 35 is located on the second terrace of the Chattahoochee, just north of the city limits of Eufaula. This area is immediately south of 1 Br 15X1, the Shorter Mound, and is separated from it by a small stream.

Preliminary surveys had indicated that 1 Br 35 was primarily a large Historic village. Documentary research, however, produced no specific evidence of an eighteenth century town in this location. The De Crenay Map of 1733 does show a Hitchiti village named Tamatle in the general region.

Investigations at the site began during the last part of the 1960 season. Because a return to the site at a later date was not anticipated, because the site was so large, and because of the nature of the remains, a decision was made to use power machinery to facilitate the excavations.

EXCAVATION

The initial operation at 1 Br 35 was the excavation of a test trench 175 feet long by 10 feet wide. The long axis of the trench was oriented east-west. This unit was excavated by 4-inch levels to sterile subsoil. A road patrol was then engaged to strip the plow zone from four long trenches, each running parallel to the test trench. These trenches measured 5 to 7 feet in width and varied in length from 500 to 750 feet. Features uncovered by this work consisted for the most part of pits. These were staked out and excavated.

As this work progressed, several burials close together in the northernmost road patrol trench were uncovered. Excavations here were enlarged to permit more complete investigation. The southernmost trench was also enlarged in the area of Feature 47, a system of post molds.

Five units were excavated at 1 Br 35 during the summer of 1960. The four road patrol trenches were designated, from north to south, Trenches 1 through 4. The initial test trench was labeled Trench 5.

A crew returned to the site in the fall of 1961 and excavated two additional units. The first of these, Trench 6, was between Trenches 3 and 4. Though this unit was adjacent to the area where the burials had been found during the previous season, no other graves were located. Trench 6 measured 40 feet in length by 15 feet in width and was excavated by arbitrary 4-inch levels. Trench 7, totaling 1,150 square feet in area, was excavated in the same manner in the

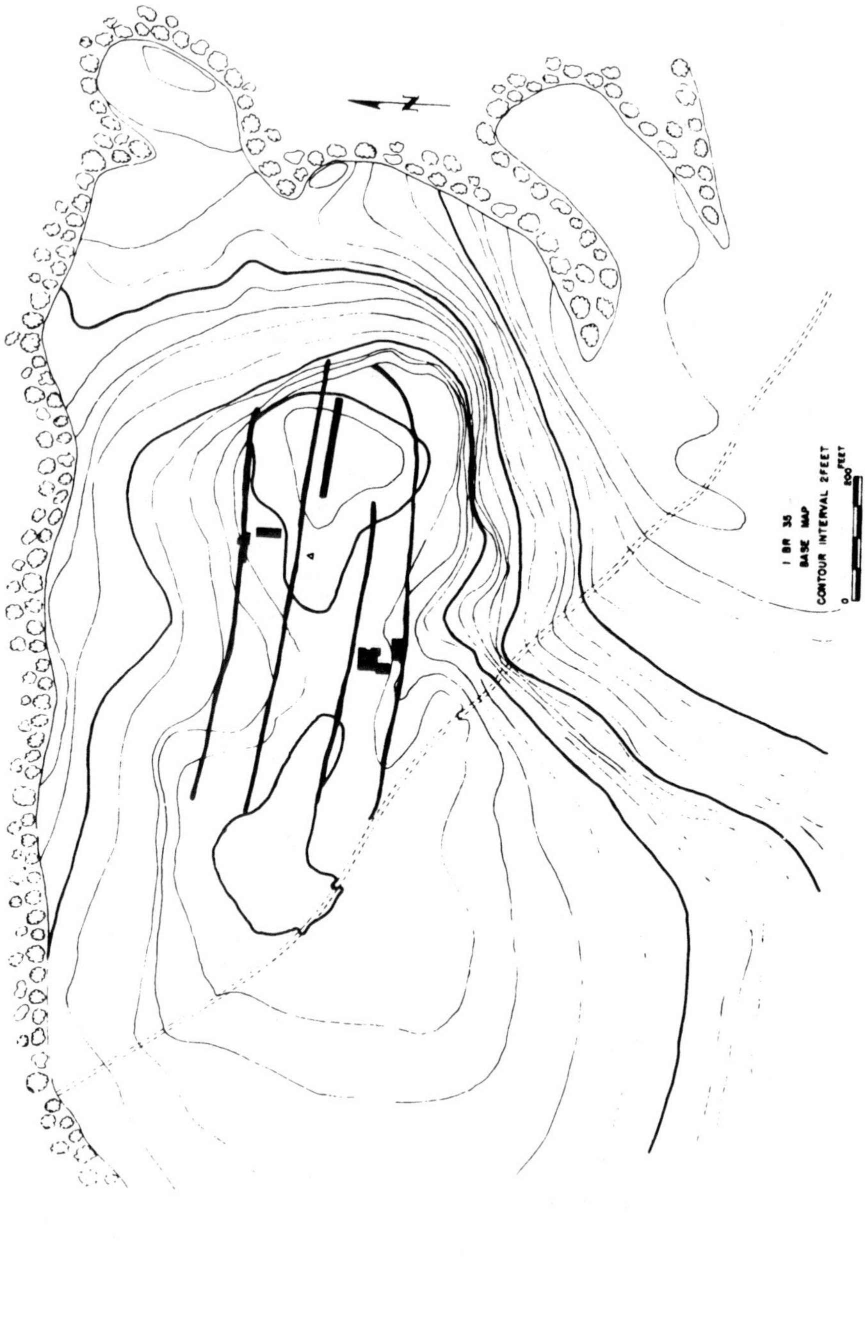

Figure 30 1 Br 35, Base Map

Figure 31 One of the Road Patrol Cuts Used to Explore 1 Br 35

Figure 32 A View of the Excavations in Trench 7, 1 Br 35

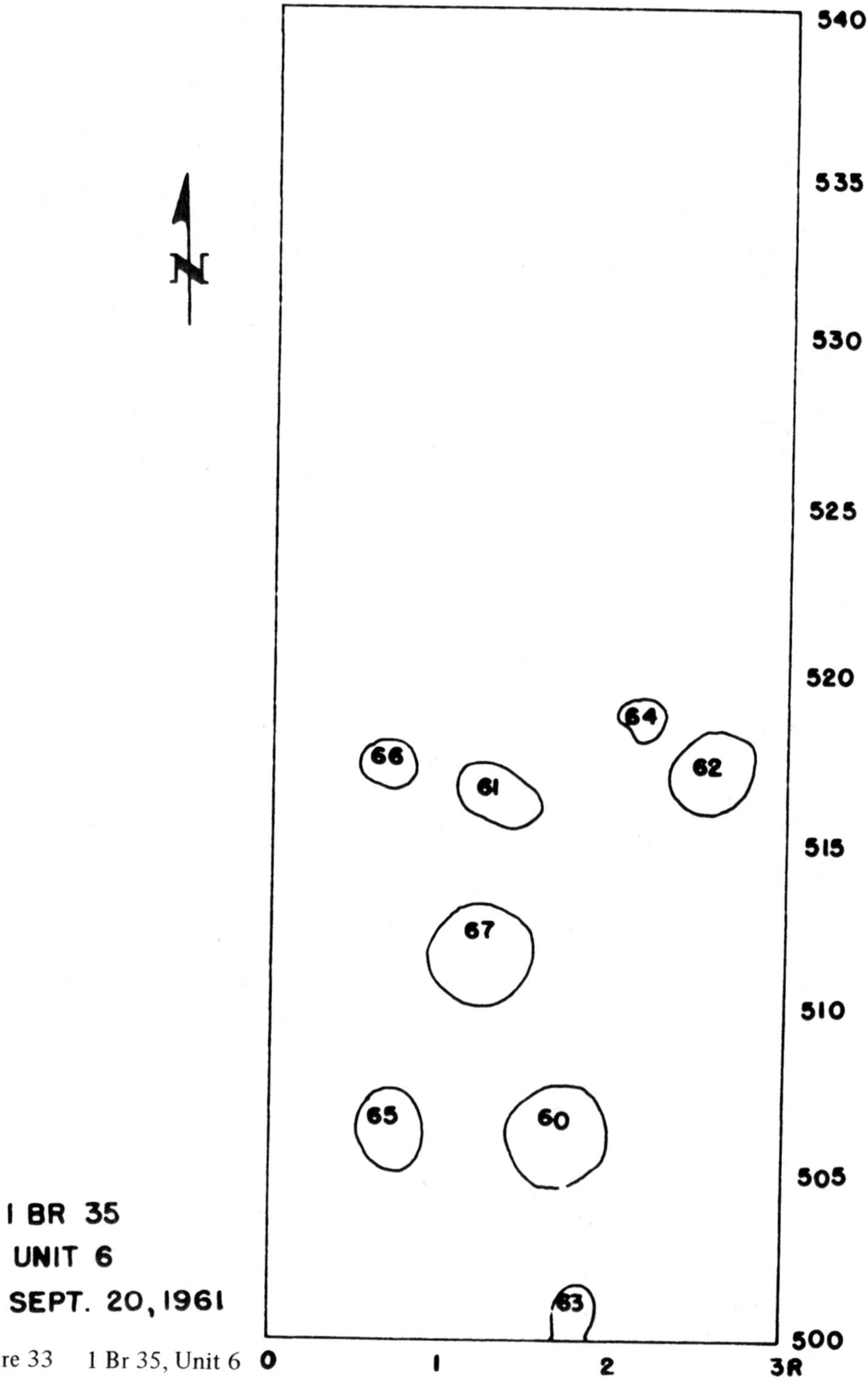

Figure 33 1 Br 35, Unit 6

area adjacent to Feature 47. This excavation was done in order to examine additional structural features.

The soil at the site consisted of a red, sandy loam. The site was in pasture for five years previous to the excavation, although crops had been harvested earlier and the soil profile showed a clear, 6-8-inch plow zone.

BURIALS

Three burials were found at 1 Br 35. One of these was interred in a fully flexed position; the remaining two, both children, were partly flexed. The exact outlines in each case were difficult to determine, but an oval shape for all of the graves was indicated. The preservation of the bone at the site was poor; all of the burials were in a fragmentary condition.

Pottery fragments were collected from the pits of Burials 2 and 3. These materials are tabulated on the chart showing the ceramic contents of burials and refuse pits.

Burial 1 was found fully flexed in a pit 21 inches deep. The burial, unaccompanied by grave goods, was that of a mature adult.

Burial 2 was interred in a partly flexed position. The pit reached a depth of 14 inches below the surface. The burial was the remains of a child and was accompanied by a number of artifacts. A small brass button was found at the left knee and a large, blue glass bead was found at the right knee. Other glass beads were around the neck and scattered throughout the burial. West of the burial, two brass fishhooks and an iron "strap" had been placed. An iron spike was found near the right humerus.

Burial 3 contained the largest amount of grave goods. This burial was that of a child placed in a partly flexed position. The grave was oval in shape and 12 inches deep. A "squat wine bottle" of a style generally considered to date from the middle of the seventeenth century was found east of the burial's skull. Four brass bracelets were found where the right arm presumably had lain and four more brass and one iron bracelet lay over the pelvic area. Glass beads were found around the neck of the burial and also scattered over the torso. A brass-plated, iron cow bell had been placed above the knees of the burial.

FEATURES

Several types of archaeological features other than burials were encountered at the site. Many of these—tree stump systems, animal burrows, and small, irregular depressions filled with surface debris—are of little interest. Most of the features created by Indian activity at the site were medium-sized, basin-shaped refuse pits. Forty of these were excavated. The largest was Feature 51-A, shown in Figure 37. It measured almost 6 feet in diameter and 32 inches in depth. The average dimensions of refuse pits at the site, however, were about 2.5 feet in diameter and 1.5 feet in depth. The pits contained sherds, animal bone, and various artifacts; most included trade materials.

Twelve small pits filled with charred corncobs were excavated at 1 Br 35.

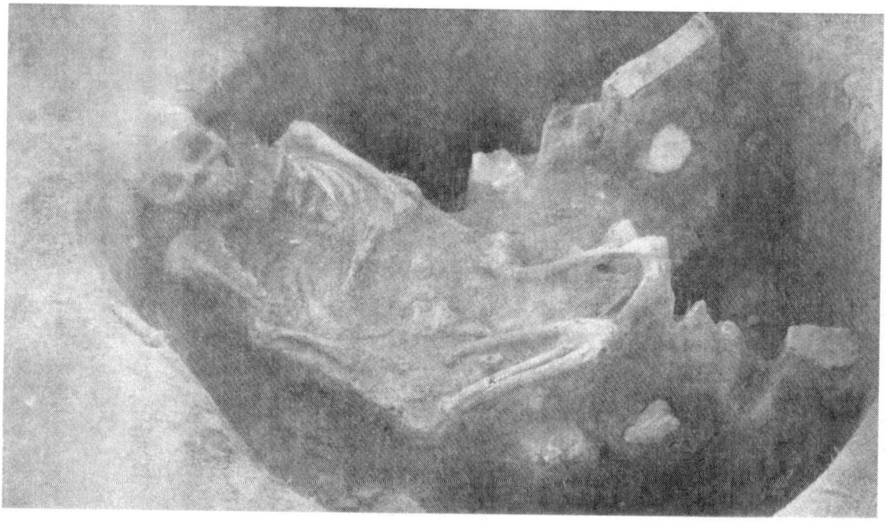

Figure 34 Burial No. 2 at 1 Br 35

Figure 35 Burial No. 3 at 1 Br 35

These pits were seldom more than 1.5 feet in diameter. In several cases they were only inches deep. Feature 2-A, one of the larger "corncob holes," is illustrated in Figure 36.

Feature 33 was the only cob-filled pit to contain ceramic fragments.

Four fire hearths were uncovered by the work at 1 Br 35. These consisted of small, basin-shaped depressions filled with fire-hardened clay, charcoal, and rocks.

A large number of scattered post molds were mapped in the course of the excavation at the site. In only two areas did these posts seem to form a pattern. Feature 47 may be a post mold combination–the remains of a circular structure about 25 to 30 feet in diameter. The post molds forming this feature were represented by bluish-gray, post-mold-shaped clay deposits in the underlying red sand of the site. These were single posts, not placed in trenches. Only a relatively small amount of wattle and daub was present, nor could a packed clay floor be detected.

Six intrusive features were associated with the post mold pattern. Feature 47-A was the remains of a tree stump in the east-central part of the enlarged excavation. Feature 47-B was a small, oval-shaped post mold 17 inches long, 11 inches wide, and 18.5 inches deep. Feature 47-C was a circular pit 15 inches in diameter and 11 inches deep. It was filled with a number of smooth, round rocks. This feature is shown in Figure 38. Feature 47-D was a deep, narrow depression, probably a post mold. Features 47-E and 47-F are also the remains of stumps. The floor plan of Feature 47 is shown in Figure 40.

A second possible post mold pattern was uncovered in Trench 7. Here an alignment of a dozen or more posts in a southwest-northeast direction was indicated (see Figure 43). Here, again, the posts were single and not placed in a footing trench.

POTTERY STUDY

Pottery was abundant at the site; a large collection was taken from both the surface and the excavated areas. Surface sherds as well as material turned up by the road patrol in the excavation of Trenches 1 through 4 form a general collection. Other material was excavated by 4-inch levels. No significant change in the frequency of ceramic types was indicated by the depth distribution study.

Soon after the investigation of 1 Br 35 began, it became apparent that the bulk of the pottery from the site was identical to the Ocmulgee Fields complex ceramics excavated at Ocmulgee Old Fields, the Lawson Fields site, and associated with numerous other Historic Lower Creek remains. The three major decorated types of the complex, Ocmulgee Fields Incised, Chattahoochee Brushed, and Kasita Red Filmed, were all present. Two varieties of plainware–one well smoothed, the other coarse and gritty–were also common. All of the sherds found at the site are classified in Table 19.

Seven pottery types form the Ocmulgee Fields ceramic complex at 1 Br 35. Only three means of decoration–brushing, incising, and red filming–were used. Short descriptions of the sherd types from the site follow.

TABLE 19

1 Br 35 – Sherd Count

	Number	Percent
Ocmulgee Fields Types		
Ocmulgee Fields Incised	286	4.717
Chattahoochee Brushed	2,713	44.754
Kasita Red Filmed	35	0.577
Walnut Roughened	102	1.683
Ocmulgee Fields Plainware		
Coarse Tempered Plainware	1,748	28.835
Fine Tempered Plainware	874	14.418
Plain Shell Tempered	200	3.299
Miscellaneous Types		
Stallings Plain	7	0.115
Cartersville Check Stamped	1	0.016
Lamar Complicated Stamped	49	0.808
Residual	47	0.775
Site Total	6,062	100.000

Brushed Types

Chattahoochee Brushed (Bullen 1950). This ware is characterized by brushed, stippled, and cob marked surfaces applied on round-bodied bowls as well as on the lower portions of incised casuelas. The first form was by far the most common. Usually the neck of a vessel of this type is left unbrushed for a short distance below the rim, but often this area is decorated with vertical or diagonal, incised, parallel lines. A notched or sometimes a pinched fillet is usually found just below the lip.

Walnut Roughened (Haag 1940a). A few shell tempered, brushed sherds were found at the site. No rims were included in the collection.

Incised Sherds

Ocmulgee Fields Incised (Jennings and Fairbanks 1939). An incised band above the shoulder of casuela bowls was the second most frequent technique used at the site for the alteration of the surface of vessels. The portion of the vessel below the shoulder was usually plain or, infrequently, brushed. The lips of these

Figure 36 Feature No. 2-A, 1 Br 35

Figure 37 Feature No. 51-A, 1 Br 35

Figure 38 Feature No. 47, 1 Br 35, Viewed from North

Figure 39 Feature No. 47, 1 Br 35, Viewed from West

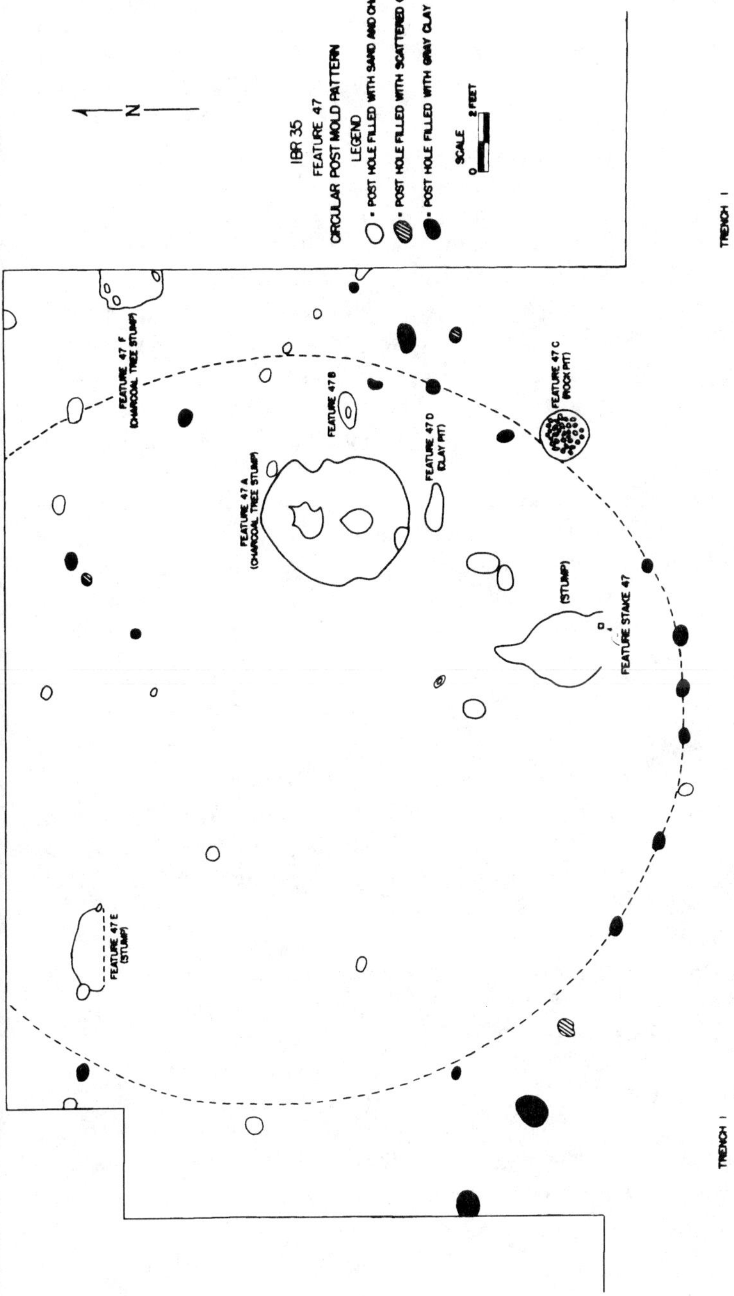

Figure 40 Feature No. 47, 1 Br 35, Post Mold Pattern

vessels were usually extruded or thickened, but simple rounded and flattened lips also occurred. All of the incised sherds were either tempered with fine sand or appeared to be temperless.

Red Filmed Sherds

Kasita Red Filmed (Haag 1940b). A small number of red filmed sherds appeared at the site. Usually the red filming was applied in zones between incised lines. The undecorated ware is most often cream-colored, but several black sherds were red filmed in the same manner.

Ocmulgee Fields Plainware

Two grades of plainware were distinguishable at 1 Br 35. The first of these was coarse, white or gray-colored, and similar in ware characteristics to the bulk of the Chattahoochee Brushed sherds. These plain sherds are thought to represent unbrushed portions of Chattahoochee Brushed vessels. Only a very few rims were classified in the Chattahoochee Brushed type; the majority of the small, coarse plainware rims found at the site probably adorned brushed vessels. Most of these coarse, plain rims possessed either a pinched rim fillet or, more frequently, a punctated fillet.

A smooth grade of plainware, similar in ware characteristics to the Ocmulgee Fields Incised type, was also found at the site. The bulk of these sherds were probably the basal portions of Ocmulgee Fields Incised vessels. A few casuela bowl fragments with no decoration above the shoulder and with plain bases were found. One smooth, plain sherd was fashioned into a crude pottery discoidal.

Miscellaneous

Several earlier ceramic types were also present at 1 Br 35. A small number of fiber tempered sherds, classified as Stallings Plain, undoubtedly represent the earliest pottery at the site. Cartersville Check Stamped and Early Swift Creek Complicated Stamped pottery are other minority wares, related to the Woodland components at the nearby 1 Br 15.

A number of Lamar Complicated Stamped sherds appeared at the site. Two varieties of this type were isolated: Sub-Type A, characterized by sloppy, smoothed-over, complicated stamping; and Sub-Type B, characterized by finer tempering materials and clearly impressed, complicated stamping.

Differences in areal distribution of ceramic types were tested in two ways. A small collection of surface and plow zone material was saved from east-west segments of the road patrol trenches. Material from Trenches 6 and 7 was compared, to test variation from north to south. Each of these separate units was compared with the others but no significant change in areal distribution was detected. Strong contrast was found when 1 Br 35 pottery was compared with the sherds from 1 Br 15, a site only a few hundred feet to the north. Brushed

Figure 41 Kasita Red Filmed and Ocmulgee Fields Incised, 1 Br 35

Figure 42 Sand Tempered Plain Vessel, 1 Br 35

wares formed only a small minority of the sherds at 1 Br 15, and check stamped pots with tetrapodal supports were the most common ceramic form.

CHIPPED STONE ARTIFACTS

Very few chipped stone artifacts or flint chips were collected at 1 Br 35. Most of the flint tools were probably not related to the Historic component. Two Mississippian Triangular (Madison) points and three Mississippian Triangular point bases were collected, as well as five large, stamped projectile points, one Benton Broad Stemmed point, two Elora serrated-like points, and the basal halves of two Savannah River points. These point types more closely resemble Archaic styles than Historic or protohistoric. Two large flint flakes showed evidence of having been used as scraping or cutting tools. Both of these were chipped unifacially.

Three uniface side and end scrapers were found at the site. Two of these are the size of gun flints and may well have been employed for this purpose, though the patinated appearance of the chert seemed to indicate a far greater age than this use would suggest.

PIPES

Seven pipe fragments of aboriginal manufacture were found at 1 Br 35. One fragment, the only ground stone artifact from the site, was the stem of a steatite pipe. The other five stem fragments were made of pottery. The stems of four of these were circular in cross section; the other was square. A single pipe bowl fragment was found.

COMPLEX OF TRADE MATERIALS

A large number of items of European manufacture appeared at the site. Many of these were clearly associated with Indian-made artifacts in refuse pits and burials. These trade objects included eight classes of artifacts: (1) glass bottles and bottle fragments; (2) claypipe fragments; (3) gun parts, including gun flints and shot; (4) iron tools, including knife blades and a pair of scissors; (5) rolled sheet brass beads, brass buttons, and other brass and iron ornaments; (6) two bells; (7) glass beads; and (8) brass fishhooks.

Glass Bottles. The single whole bottle found at the site was associated with Burial 3. It is almost identical to a bottle found at the Childersburg site, which was dated around 1730-1740 by Malcolm Watkins of the United States National Museum (DeJarnette and Hansen 1960). Two necks and a bottle base resembling corresponding sections of the whole bottle were also found. In addition, 82 green glass wine bottle fragments were found. Many of these were heavily patinated. All exhibited air bubbles within the glass. At least one showed secondary chippage and other signs of use as a scraper.

Claypipe Fragments. Kaolin pipe stem and bowl fragments were common on

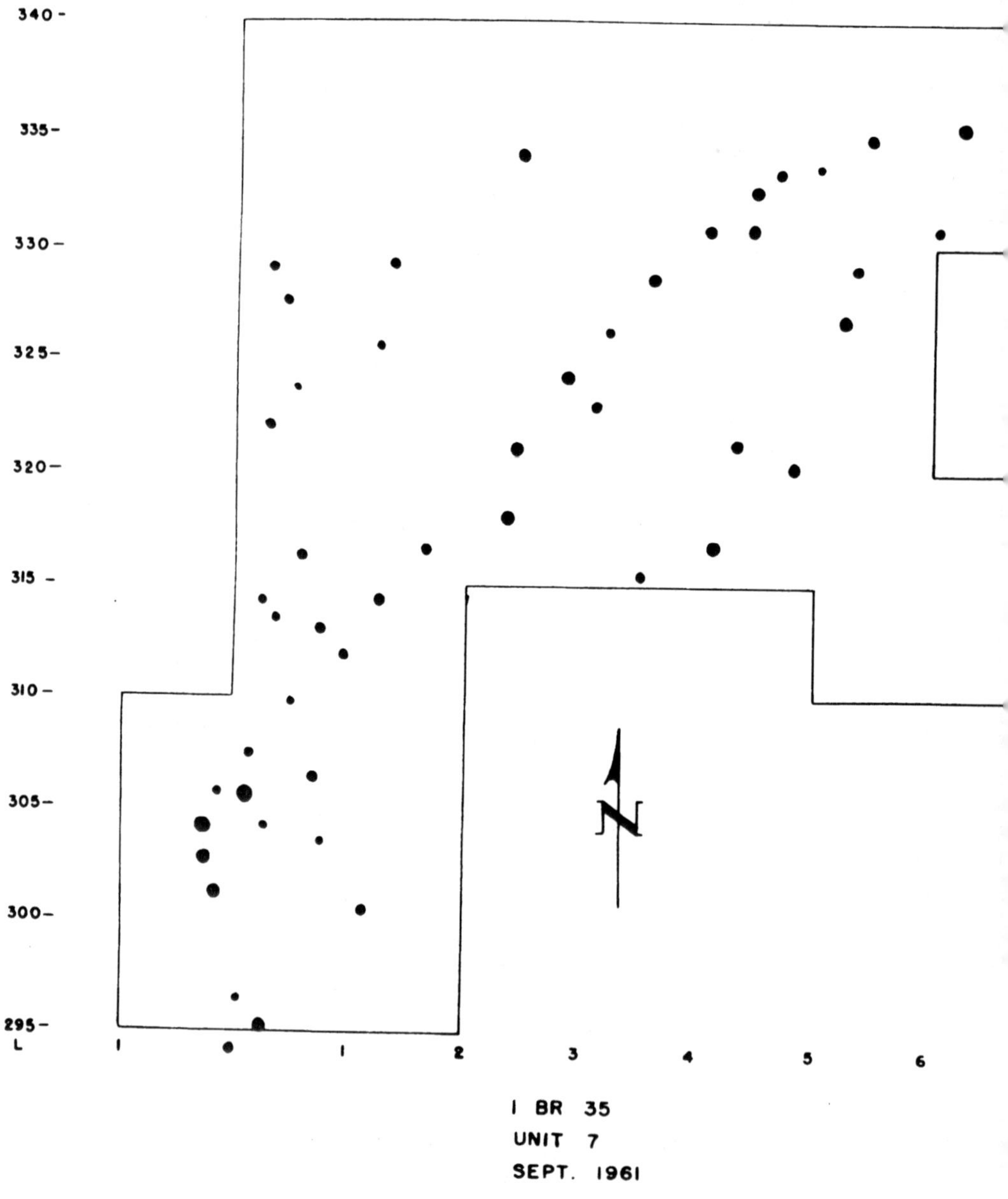

Figure 43 1 Br 35, Unit 7

both the surface and plow zone of the site and in refuse pits. Ninety-two stem and 35 bowl fragments were collected. The bore diameter of each of the stems was measured. Of the total sample of 92, 17.4 percent were 4/64-inch; 78.3 percent were 5/65-inch; and 4.3 percent were 6/64-inch. The high frequency of 5/64-inch bore diameters indicated a 1710-1750 date for the pipes. The Binford formula for the average deposition gives 1773, a slightly later date. All of the bowl pieces were carefully examined, but no maker's marks were noted.

Gun Parts. Two frizzens, a gun thimble, three gun barrel sections, a brass trigger guard, gun flints, and five balls of lead shot uncovered at the site were evidence of the use of firearms. The barrel sections were 6 to 8 inches long and flattened at one end. Hamilton (1960: 126) documents the use of metal from gun barrels as fleshers; these artifacts were probably used in this manner. The gun flints showed a great deal of wear from use. As a group, they seemed to be of Indian rather than European manufacture.

Iron Tools. Three iron fragments, probably representing rusted knives, and a pair of rusted scissors were found in the plow zone and with the pits and burials at the site. Nails of all sizes were common artifacts. Seventeen of these were found, several in refuse pits. Other pieces of metal are more than likely modern.

Metal Ornaments. Eight brass bracelets and one iron bracelet were found with Burial 3. A brass button, similar to those found at the Childersburg site (DeJarnette and Hansen 1960), was associated with Burial 2. The button was of the high dome type, a shape characteristic of Queen Anne (1702-1714) furniture and hardware. The brass buttons from Childersburg were placed in the eighteenth century by Watkins.

Bells. Two bells were recovered at 1 Br 35. One was a small, light "hawk bell" and the other was the cow bell associated with Burial 3. The cow bell was copper-plated.

Glass Beads. Eleven bead types were found associated with the burials or in other contexts. The bead types are listed below with an indication of their occurrence.

BEAD CLASSIFICATION FOR 1 Br 35

Childersburg Transparent Spherical. Sixty-eight beads of this type were found at 1 Br 35 in direct association with Burials 2 and 3 and Pit D. The beads found with Burial 3 were in position in the neck region, which would suggest that they were worn as a necklace rather than decoration on the clothing. All of the above beads were of the seed variety.

Tallasseehatchee Transparent Decahedral. Only one of this type was found at

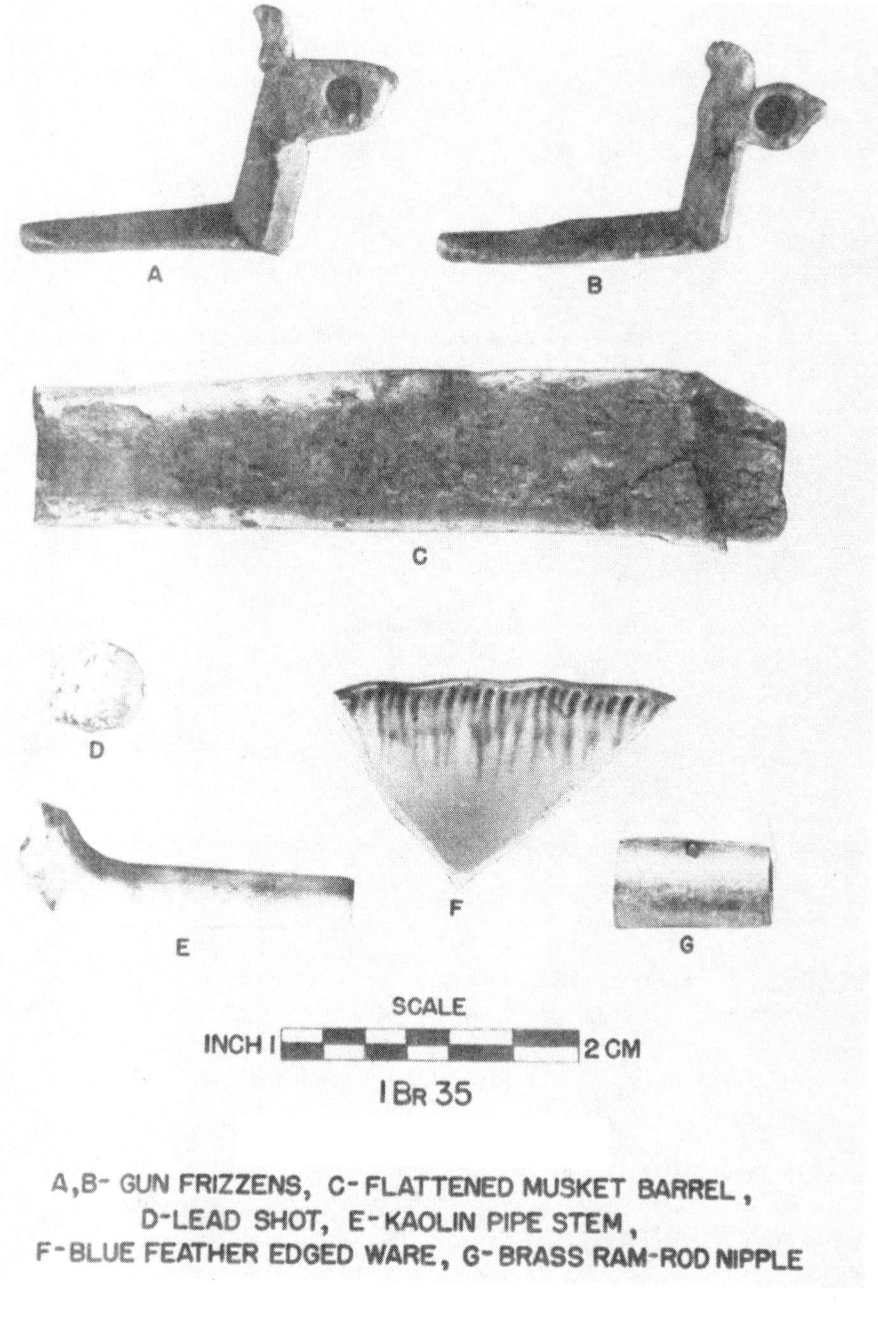

Figure 44 1 Br 35, European Trade Material

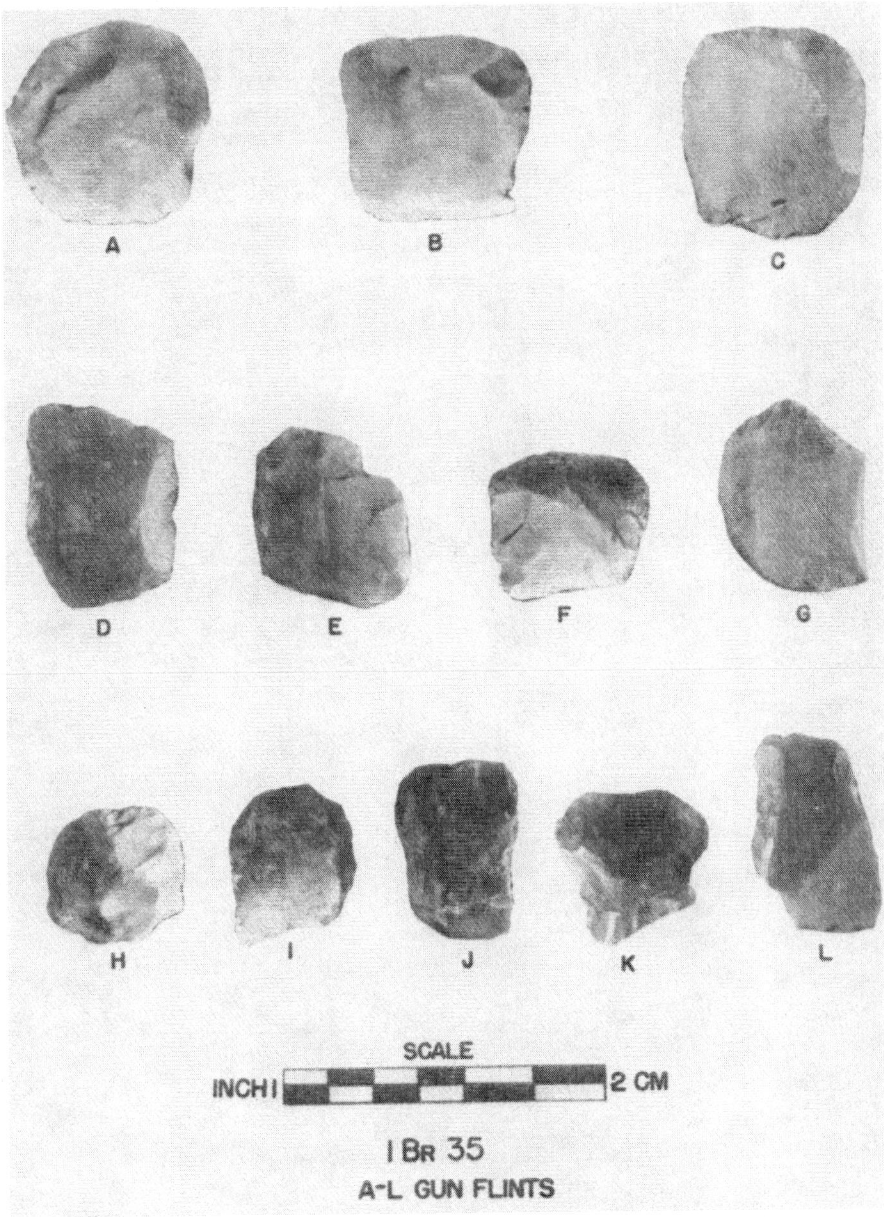

Figure 45 1 Br 35, Gun Flints

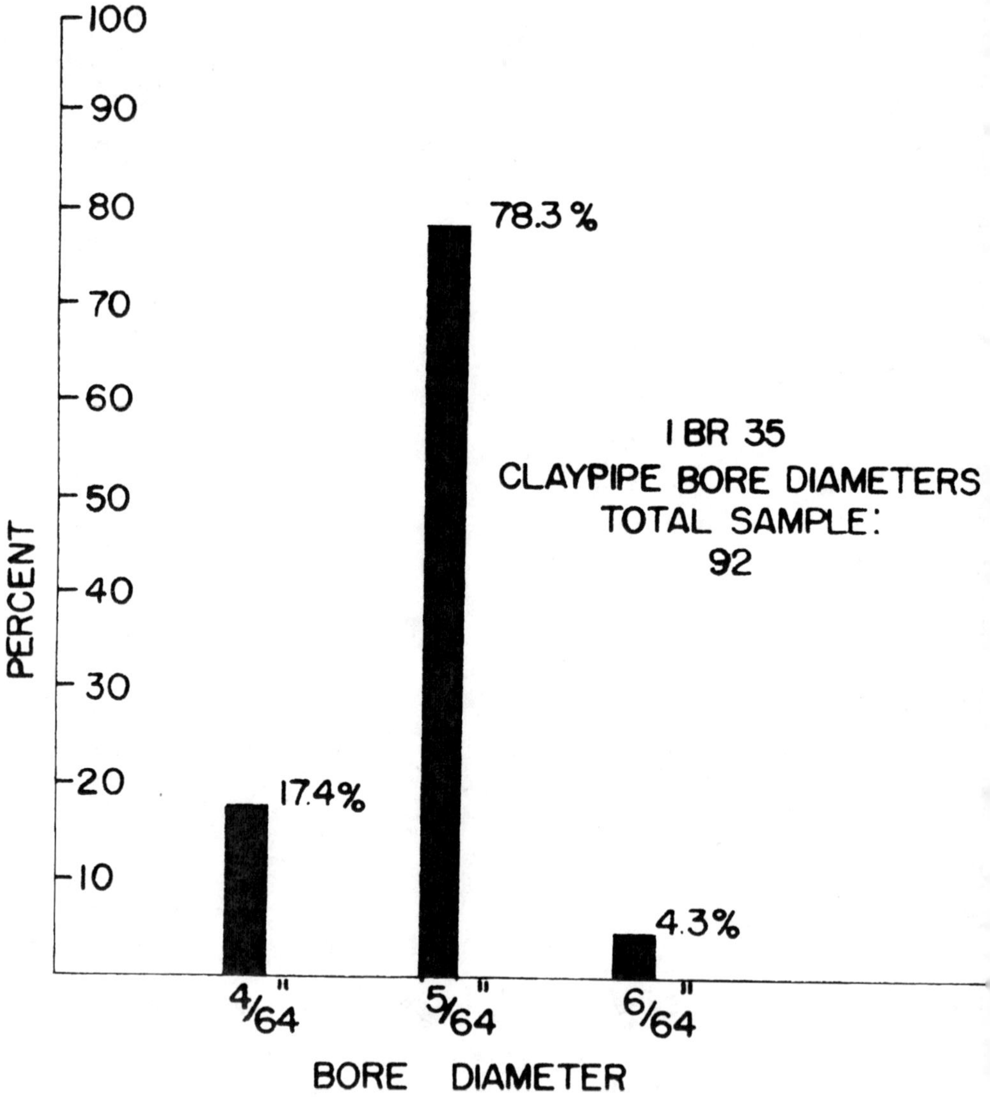

Figure 46 1 Br 35, Claypipe Bore Diameters

1 Br 35. It was in a poor, badly eroded condition, and was encountered in Feature 14. It was pale blue in color, transparent, and measured approximately ¼ inch in diameter.

Childersburg Transparent Dark Blue Oval. Two of these were found at 1 Br 35 associated with Burial 2, just north of the right leg. This type of bead is rather large and dark blue in color.

Tallasseehatchee Transparent Yellow. This type was not associated with any of the burials but was found at Level 1, Trench 3. Only a fragmentary portion, approximately ¾ of an inch in diameter and of a transparent lemon yellow, was recovered.

Childersburg Opaque Blue Oval. Three of these were found at site 1 Br 35, one at Level 1, Trench 2; one at Feature 4; and one at 39 E and F at the west end of Trench 2. This is a rather large, deep blue type, round or oval in shape, approximately ¾ of an inch in diameter.

Fort Moore White Striped Spherical. These were all of the seed type or slightly larger. Each displayed a black base and four white stripes. (The stripes did not run at an angle.) Two hundred and one of these were found, at the following locations: Feature 1, Trench 4; Feature 59, Pit D, Trench 4; neck region and elsewhere in Burial 2; and Burial 2, Pit D.

Georgia White Cylindrical. Six beads were recovered at Level 1, Trench 4, and Level 2, 291-5. They varied in length but all were chalky white in color, and approximately ¼ inch in diameter.

Georgia Translucent Blue Cylindrical. A single bead was found associated with Burial 2. It was dark blue in color, translucent, and approximately ¼ inch in diameter.

Georgia Milk Oval. One bead of this type was found at 1 Br 35. It occurred on the surface of Trench 4. It was milky white in color, barrel-shaped, 1 inch long and ½ inch in diameter.

Fort Moore Transparent Purple. Eleven beads of this type were found at the site in the following areas: Burial 2, neck region; Burial 3; Burial 2, Pit D; and Burial 2, Level 1, Trench 4. They were barrel-shaped, of a clear purple color, approximately ¼ inch in length and approximately 1/8 inch in diameter.

Alabama White and Black Seed. Many beads of this type occurred at the site, associated primarily with Burials 2 and 3. A total of 67 black and 555 white beads were located at the following areas: Burial 2, neck region, Level 2; Level 1,

Trench 3; and Burial 3, Level 3; Feature 59, Trench 4; Level 1, Trench 4; and within the post hole in Feature 59. Again the presence of some beads in the region of the neck suggests that they were worn around the neck rather than on the clothing.

FISHHOOKS

Two small brass fishhooks were found with Burial 2. A third brass fishhook, slightly larger, was taken from Feature 38, a refuse pit.

MISCELLANEOUS

Several transfer-printed, hand-painted, dendritic and feather edge sherds were collected from the surface and plow zone of the site. This complex of chinawares probably dates circa 1830. No chinaware fragments from this complex were associated with the Indian material in pits or features. It is therefore concluded that the Creek occupation of the site ended before that date.

CONCLUSIONS

1 Br 35 is principally a component of the Ocmulgee Fields phase. It was occupied by the Creeks during the mid eighteenth century, if the date inferred from the trade materials (primarily the claypipes and bottle) is correct. The only town known from historical records to be in this general area at that time was Tamatle. Site 1 Br 35 may have been part of this village.

By the year 1750, the Creeks had undergone considerable culture change due to influences from Europeans, especially traders. Evidence of this acculturation is the scarcity of flint chips and fragments at site 1 Br 35. Flint debris occurred in large amounts at the adjacent site, 1 Br 15, but was almost entirely absent at 1 Br 35. Fragments of gun parts, however, were not at all rare at 1 Br 35.

The extent of the acculturation becomes even more apparent when the burial furniture is examined. Although only two burials accompanied by grave goods occurred in the excavated area, many artifacts were found in association with these two burials. None of these artifacts was of Indian manufacture.

Resistance to culture change, however, can be seen in the ceramics of the Ocmulgee Fields phase (Fairbanks 1961). That the change did take place is easily demonstrated by both the archaeology and the ethnohistory of the area, yet Creek pottery found at 1 Br 35 is basically of the same complex as that found at Ocmulgee Fields, which dates considerably earlier.

1 Br 37, The Fontaine Reeves Site

1 Br 37 is located 1300 feet northeast of 1 Br 15. The site, situated on a bluff that parallels the Chattahoochee River at a distance of about a half mile from its bank, is separated from 1 Br 15 by a deep gully and river branch. The general area of the site is just north of the city limits of Eufaula (Figure 20).

Pottery fragments were lightly scattered over this field and slightly concentrated at its southern margins. A formal grid system was laid out in the south central part of the site. Four small rectangular units were excavated using this grid system for horizontal control. The areas investigated proved to contain only a small amount of cultural material. Seven test pits were then excavated in the area slightly to the northeast of the initial trenches. These also proved to be largely devoid of aboriginal remains. No refuse pits or other archaeological features were located by the digging described above. Had any features been present, they would have been easily distinguished by their intrusion in the hard clay subsoil.

The material collected at the site consisted of a ceramic collection and a small number of projectile points and other stone artifacts.

CERAMIC ANALYSIS

Most of the sherds collected at the site were undecorated body sherds. The largest group of these plain sherds was labeled "Coarse Plain." Most of the sherds placed in this category were body sherds from Lake Jackson Plain vessels or unmarked body fragments of Chattahoochee Brushed pots. The sherds labeled "Smooth Plain" were similar to the fine grade of plainware from site 1 Br 35. The most easily recognized category of plain sherds was typed "Micaceous Plainware." This ware was tempered with a combination of fine sand and a quantity of mica. "Micaceous Plain" sherds were identical in ware characteristics to the Cartersville Check Stamped sherds found at 1 Br 15. The Weeden Island Plain and Carrabelle Punctated sherds found on the site were also characterized by the same type of temper.

The following table summarizes the pottery found at the site. No significant difference in type frequencies was found with depth.

NON-CERAMIC ARTIFACTS

A gun flint of intermediate size made from waxy, gray flint was collected on the surface of 1 Br 37, and two green glass wine bottle fragments were found in the excavations. These artifacts were probably related to the Ocmulgee Fields pottery at the site.

Four types of "Archaic style" artifacts were collected from the site: (1) a straight-stemmed projectile point, Elora-like and similar in appearance to many found at 1 Ru 28; (2) half of a sandstone atlatl weight, diamond-shaped in cross section; (3) two drills—one expanded base, the other a reworked, stemmed point; (4) two steatite sherds.

Four small, stemmed projectile points and a stemmed point fragment similar to the Ebenezer type were taken from the site. One Mississippian triangular (Hamilton) point was also found.

SUMMARY

The earliest occupation of site 1 Br 37 seems to have taken place during the Early Woodland and Late Archaic periods. Harold Huscher reported finding fiber tempered pottery fragments at the site, and several artifacts found in the excava-

TABLE 20

1 Br 37 Pottery Summary

Types		Number	Percent
Ocmulgee Fields			
Chattahoochee Brushed		172	14.39
Walnut Roughened		1	.08
	Total	173	14.47
Lamar			
Fort Walton Incised		7	.59
Lake Jackson Plain		7	.59
	Total	14	1.18
Weeden Island			
Weeden Island Plain		2	.17
Carrabelle Punctated		3	.25
	Total	5	.42
Early Swift Creek			
Swift Creek Complicated Stamped (Early Variety)		16	1.34
	Total	16	1.34
Cartersville			
Cartersville Check Stamped		1	.08
Deptford Bold Check Stamped		1	.08
Dunlap Fabric Marked		1	.08
	Total	3	.24
Plainware			
Coarse Plainware		426	35.65
Micaceous Plainware		246	20.59
Smooth Plainware		289	24.18
Plain Shell Tempered		23	1.92
	Total	984	82.34
	Grand Total	1195	100%

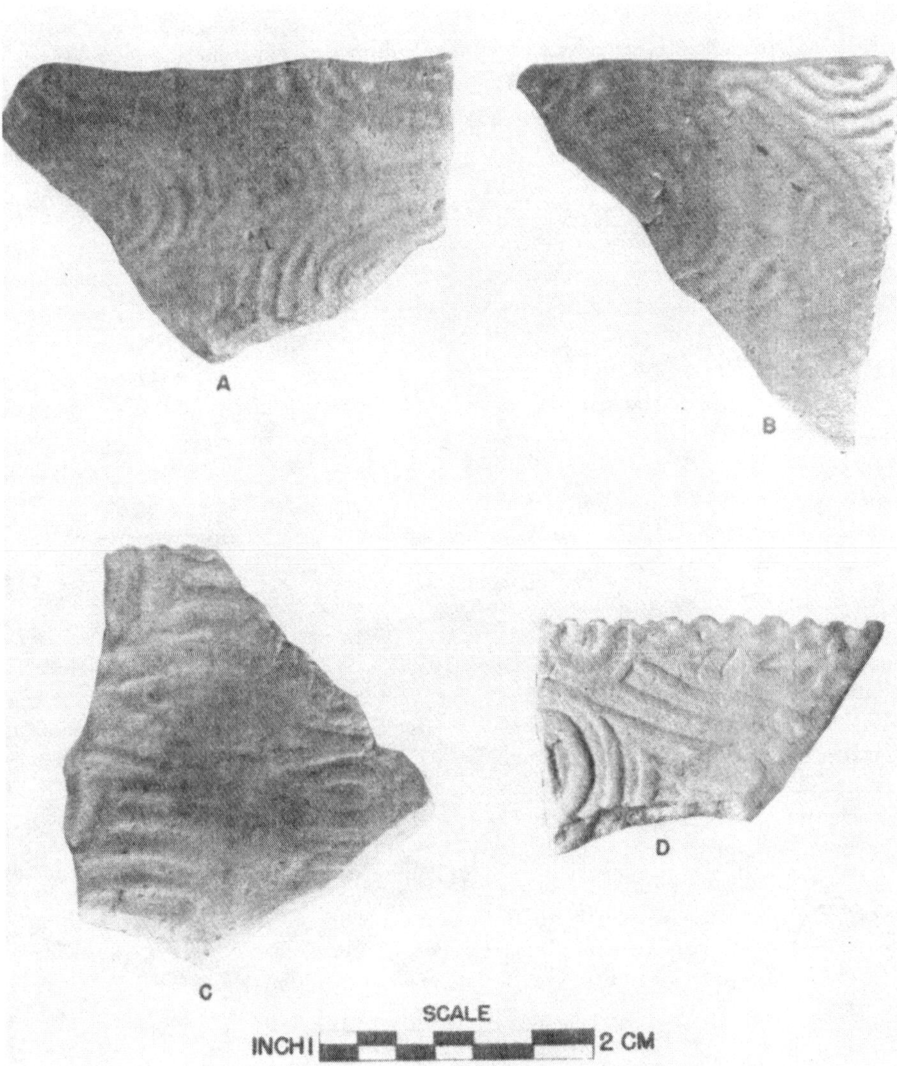

Figure 47 1 Br 37, Early Swift Creek Complicated Stamped

tions are indicative of the presence of a complex of fiber tempered pottery, stemmed projectile points, and steatite sherds, similar to the Archaic component at 1 Ru 28.

Cartersville phase sherds related to the Cartersville occupation at 1 Br 15 were present at 1 Br 37. Several Early Swift Creek phase sherds were found scattered in the various areas of the site. Other pottery fragments suggested a small Weeden Island occupation.

The Ocmulgee Fields component pottery and the related European trade artifacts form the bulk of the collection from the site. The Historic Creek Ocmulgee Fields complex, present at 1 Br 15, reaches its greatest concentration at site 1 Br 35. The 1 Br 37 Ocmulgee Fields component is probably related to the middle to late eighteenth century occupation there.

1 Ru 28, The McLendon Site

The first phase of the excavation of the McLendon site was undertaken during the summer of 1960 by the student field party of Florida State University under a cooperative agreement between The University of Alabama and Florida State University. Lewis Larson was in charge of this excavation. The results of this first excavation indicated a deep component of Archaic material that prompted more extensive excavation of the site.

A proposal for additional excavation and study was then submitted to the National Park Service, and additional funds were made available from that agency. During the summer of 1962 Edward B. Kurjack, assisted by John Cottier, supervised a field crew from The University of Alabama in further investigations of site 1 Ru 28 in order to obtain more information concerning the earlier components.

LOCATION AND PHYSICAL DESCRIPTION

The McLendon site is located 2 miles east of Cottonton, Alabama, in Russell County. It is situated on the Alabama side of the Chattahoochee River, immediately south of the mouth of Hatchechubbee Creek. The site is on the second terrace of the Chattahoochee, in a field near the sloping banks of the river. Before its acquisition by the U.S. Corps of Engineers, it was owned by J. C. McLendon, who cooperated freely with the excavators. The property is in the southeast quarter of Section 6, Range 30 East, Township 13 North.

All of the excavations were made at the extreme east margin of the field. The area between the field and both the Chattahoochee River and Hatchechubbee Creek was once heavily wooded; now, however, these woods have been cleared. The site is just above the cleared area. Site elevation ranges from 190 to 195 feet.

METHOD OF EXCAVATION

The excavations at site 1 Ru 28 during the summer of 1960 consisted of a series of 13 pits, each 10 feet square. All excavation was done by 6-inch levels, stratigraphic control being provided by balks left between adjacent squares.

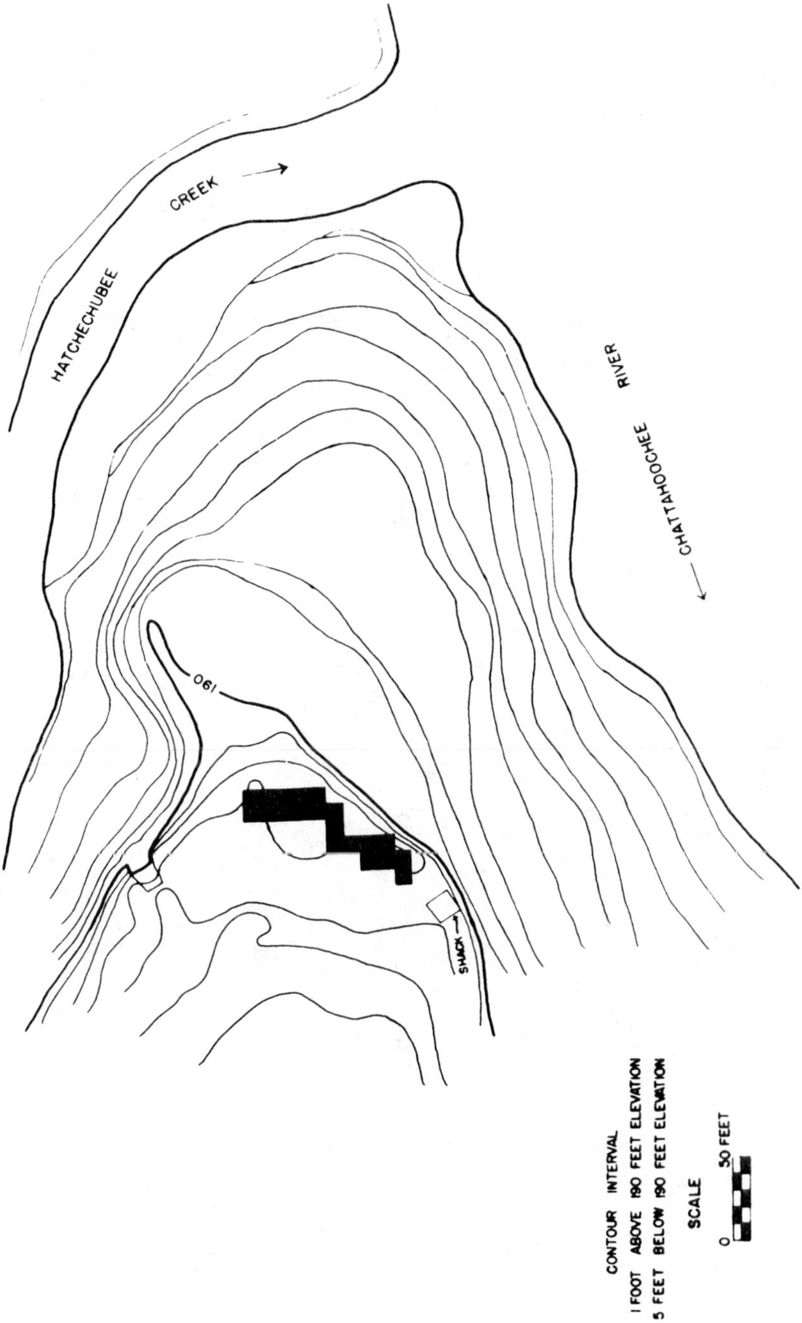

Figure 48 1 Ru 28, The McLendon Site

Figure 49 One of the Series of Test Pits Excavated During the Summer, 1960, 1 Ru 28

This work revealed a simple soil profile. The upper 2 feet of the site consisted of loosely consolidated gray-white sand. Hard, yellow-brown, sandy clay was then encountered at depths varying from 2 to 3 feet.

Most of the artifacts collected during the first season's work were taken from the upper sand layer; however a few artifacts were found deep in the clay, which at first had appeared to be sterile. The purpose of the second season's work at the site was to examine this clay layer carefully and to test for the presence of a deep component.

Two blocks were isolated by cleaning off the northernmost of the 1960 profiles and excavating two 10-by-20-foot units on two sides. The sand layer was then removed from the blocks as shown in Figures 50 and 51. Only one artifact, the tip of a projectile point, was found below the sand layer, and it was taken from the first 6 inches of clay. Most of the excavation was carried to a depth of 6 feet, although square 360L335 (designated by the coordinates of the southeast corner) was excavated to a depth of 12 feet. Six-inch levels were used during 1962 work. Screening was employed during both seasons.

FEATURES

Four special features—all fireplace areas consisting of piles of fire-cracked cobble rock—were defined at the site. Small flecks of charcoal were present in all of the features. Two fiber tempered sherds were associated with Feature 2, and a projectile point with Feature 4.

Larson's field journal contains records of numerous fired areas or hearths at various levels. Two deserve special mention. The first, found approximately 4.2 feet below the surface, was a somewhat inconsistent pattern of pebbles, many of which were fire-cracked and charred. A quantity of charred hickory-nut shells and what appeared to be a charred locust seed, as well as a large, stemmed projectile point, flint chips, and a hammerstone, were found in association.

The second hearth, encountered 6.47 feet below the surface, consisted of a few widely scattered stones. Identification as a hearth depended in part upon the presence of charred hickory-nut shells. Whether or not these shells originated at this level or were found here because of an intrusive pit from the above hearth is not certain.

POTTERY STUDY

Seven hundred and twenty-five sherds, representing at least three and possibly four recognizable components, were found at 1 Ru 28. Most of this collection consisted of Historic Creek pottery types; however, a small number of body sherds, one pinched rim fillet, and a duck's head effigy did not seem to be related to the Historic component. These artifacts are considered to be evidence of a short Fort Walton-Lamar occupation. A third group of related sherds consisted of undecorated ware with fine sand and mica temper. The earliest ceramic phase present at 1 Ru 28 was represented by a fiber tempered component.

Figure 50 Beginning the Excavation of Block 1 During the Summer, 1962, 1 Ru 28

Figure 51 Block 1, with the Clay Layer Uncovered, 1 Ru 28

Figure 52 Excavating the Sand Layer on the East Side of Block 1, 1 Ru 28

Figure 53 Block 1, Before the Excavation of the Clay Layer, 1 Ru 28

Five ceramic types and five sorting categories were used to describe the pottery collection at the site. Each of these is described below.

Ocmulgee Fields Types

Chattahoochee Brushed (Bullen 1950; Goggin 1953). Most of these sherds were tempered with coarse grit. All were heavily brushed on the outer surface. No stippled or cob marked sherds were collected at the site. All but two of the sherds were body fragments; these two were brushed rim sherds with small, pinched rim fillets. The Chattahoochee Brushed sherds included most of the pottery that could be definitely assigned to the Historic component. No Walnut Roughened sherds were collected at the site.

Kasita Red Filmed (Anon. 1940b). Three small, thin body sherds of a fine grade of Ocmulgee Fields-like pottery were zone red filmed. These sherds closely resembled Kasita Red Filmed pottery from other nearby Historic sites.

Ocmulgee Fields Incised (Jennings and Fairbanks 1939). One incised casuela fragment was found on the surface of the site.

Ocmulgee Fields Plainware. A large part of the collection from the site consisted of undecorated sherds similar in ware characteristics to either the coarse grit tempered Chattahoochee Brushed type or the more finely tempered Ocmulgee Fields Incised. These sherds were grouped into an Ocmulgee Fields Plain sorting category.

Fort Walton-Lamar Types

Lake Jackson Plain (Willey 1949). A single rim was classified as Lake Jackson Plain on the basis of its slightly finer and more uniform temper (compared to the temper of Ocmulgee Fields Plainware) as well as the size of its pinched fillet.

Plain Shell Tempered Ware. A few shell tempered body sherds were present at the site. Because of the lack of any shell tempered, brushed sherds in the Historic component, these were grouped with the Fort Walton-Lamar types.

Lamaroid Plainware. A few sherds were classified in this sorting category on the basis of their ware characteristics. Although this ware is coarse grit tempered, its temper is not quite so coarse as most of the Chattahoochee Brushed or Ocmulgee Fields Plain pottery. The sherds seemed to be fragments of Lake Jackson Plain vessels.

Fiber Tempered Types

Stallings Plain (Griffin 1943; Sears and Griffin 1950). Several fiber tempered body sherds were present at the site. None of these was decorated, and only three rims were present. Rims seemed to be straight and simple with either

Figure 54 Excavating Block 2, 1 Ru 28

Figure 55 Feature 3, One of the Fireplace Areas or Cobble Rock Piles, 1 Ru 28

rounded or slightly flattened lips. The thickness of the body sherds ranged from 0.4 cm. to 1.6 cm. and averaged approximately 0.8 cm.

The type name "Stallings Plain" is employed by David Chase in his description of Chattahoochee Valley pottery. The presence of Stallings Punctated sherds in the Chattahoochee basins provides additional justification for the use of the Stallings type name for the Chattahoochee Valley fiber tempered material.

Miscellaneous Types

Micaceous Plainware. Sherds classed in this category were tempered with fine sand and mica. The surfaces of uneroded sherds were smooth; however, few sherds remained uneroded. Evidently the pottery was not fired as "hard" as the Historic and Fort Walton-Lamar wares. The category is similar to Kellar and McMichael's Type C Plainware (1959) and Willey's Smooth Plain category (1949). Three rim sherds of this ware were collected, all having slight rim folds.

Weeden Island Zoned Red (Willey 1949). A single, large body sherd from the site was decorated by the application of red pigment in zones outlined by incised lines.

The frequency of each ceramic type at the site is indicated in Table 21. The depth distribution of the types is shown in Table 22. The relative frequency of fiber tempered wares was greater in the levels below one foot, while the Ocmulgee Fields types were found in greatest quantity near the surface. The diagram indicates an intermediate placement for the other wares.

ARTICLES OF EUROPEAN ORIGIN

A significant proportion of the artifacts found at 1 Ru 28 had been traded to the Historic Creek group that occupied the site. Seven green glass bottle fragments were on the surface and in the first two 6-inch levels of the site. Two gun flints—one almost certainly of English manufacture—were also collected on the surface. A single rolled-brass bead was found at a depth of 12 to 18 inches.

Other collected objects of Euro-American origin may have been used by others than the last group of Indians to inhabit the site. Several metal fragments, two nails, and a jew's-harp were collected in the upper levels of the site. The nails, at least, were modern. Nine chinaware fragments of various types constituted another group of artifacts probably not associated with Indian activity at the site.

CHIPPED STONE ARTIFACTS

Projectile Points. Forty-nine projectile points and fragments were recovered at 1 Ru 28. Ten of the projectile points and fragments were of quartzite; thirty-nine were of flint, including jasper, chalcedony, and chert. Only one of these indicated the later occupations of the site: this was the tip end of a Mississippian Triangular or Madison point. All the other identifiable types were of the

TABLE 21

1 Ru 28 – Pottery Count

	Number	Percent
Ocmulgee Fields		
Chattahoochee Brushed	215	29.65
Kasita Red Filmed	3	.41
Ocmulgee Fields Incised	1	.13
Ocmulgee Fields Plain	272	37.53
Total Ocmulgee Fields	491	67.72
Fort Walton-Lamar		
Lake Jackson Plain	1	.13
Plain Shell Tempered	21	2.90
Lamaroid Plainware	3	.41
Total Fort Walton-Lamar	25	3.44
Fiber Tempered		
Stallings Plain	80	11.03
Total Fiber Tempered	80	11.03
Miscellaneous		
Micaceous Plainware	93	12.83
Weeden Island Zoned Red	1	.13
Unclassified	35	4.83
Total Miscellaneous Ware	129	17.79
Site Total	725	100.00

TABLE 22

Occurrence of Pottery, 1 Ru 28

Types Level	0"-6"	6"-12"	12"-18"	18"-24"	24"-30"	Surface	Debris	Feature 2	Total
Ocmulgee Fields									
Chattahoochee Brushed	128	32	14	1	8	15	17		215
Kasita Red Filmed	3								3
Ocmulgee Fields Plain	111	68	29	9	10	15	30		272
Ocmulgee Fields Incised							1		1
Fort Walton-Lamar									
Lake Jackson Plain			1						1
Lamaroid Plainware	1	2							3
Plain Shell Tempered	16	1				2	2		21
Fiber Tempered									
Stallings Plain	26	8	23	3	1	2	15	2	80
Miscellaneous									
Micaceous Plain	46	23	18	2			4		93
Weeden Island Zone Red				1					1
Unclassified	24	2	3	2		3	1		35
Total	355	136	88	18	19	37	70	2	725

stemmed or rounded base types that have been recognized as Archaic or Early Woodland.

A single projectile point chipped from quartzite was similar to points illustrated by Caldwell (1958: 7) as representative of the Old Quartz industry. The general shape of this point and of those illustrated by Caldwell related them to Coe's Morrow Mountain types from the North Carolina Piedmont (n.d.).

Four examples (Fs 100-1, 107-1, 109-1, 147-1), also made of quartzite, were classified as Provisional Type 12 (rounded base). All were recovered from Levels 2, 4, and 5. Other quartzite projectile points and scrapers were present, but the label "Old Quartz" may not be fitting. Kellar and McMichael (1959) note that ". . . quartz and quartzite were used by most, if not all, occupants of the Oliver basin, and attempting to segregate a 'culture' using quartz exclusively is dubious procedure." This situation seemed to be the case at 1 Ru 28, where form appeared to be more important than material. An Appalachian Stemmed (Fs 100-2) projectile point, made of pink jasper but identical in form to quartzite examples from North Carolina, Georgia, and Washington County, Alabama, was recovered; and, conversely, a later, Adena type projectile point made of quartzite was found.

Three specimens of the Savannah River type were found, two in Levels 3 and 4 and the third from debris where the exact stratum could not be ascertained. Other types found were the Adena (Fs 101-2) on the surface, an Adena Narrow Stemmed (Fs 113-1) from Level 4, a Pickwick Expanded Barb (Fs 114-1) also from Level 4, and a Maples Broad Stemmed from Level 7.

Uniface Scrapers. A long, "snub-nosed," uniface end scraper was collected at the site. One other tool–a long, narrow, thin, flake knife–exhibited secondary unifacial chippage on one side. Three large flint flakes showed evidence of having been used as scrapers and had not been further shaped or sharpened by secondary chippage.

Biface Quartzite Core Scrapers. Several artifacts were shaped by bifacial chipping of quartzite cobble rocks. These artifacts were barely large enough to have been used as axes or choppers; more likely they were employed as scrapers.

Biface Scrapers. Artifacts of this type would be called "crude tools" in the Tennessee Valley area nomenclature (Cambron and Hulse 1960). Several small quartzite scrapers of "blanks" were examined separately, in order to assess the cultural significance of quartzite in the area. A single example of a stemmed scraper, a Savannah River projectile point reworked to a scraper (Fs 117-1), was collected.

GROUND STONE ARTIFACTS

Atlatl Weight. A single bannerstone or atlatl weight was recovered.

Metates. Several metate or milling stone fragments were present. These artifacts, considered a diagnostic Archaic characteristic, are evidence of seed gathering.

Hammerstones. Many of the cobble rocks that were abundant in all levels of the upper sand layer of the site were battered at the ends. A few of these may have been used by the Indians; most of them were probably naturally shaped. One pitted hammerstone was collected.

Stone Vessel Fragments. Nine steatite and three sandstone sherds were collected at 1 Ru 28.

Flaked and Ground Axe. The axe was flaked from hard limestone or dolomite.

SUMMARY

Three principal components are found at 1 Ru 28. A Historic Creek-Ocmulgee Fields occupation is represented by a large part of the ceramic collection, and by glass bottle fragments and other trade artifacts. This component is presumed to be part of the remains of an eighteenth century village. Documentary evidence places Sawokli town in the general vicinity, and 1 Ru 28 may have been part of this village, but compared to the other large, rich, Ocmulgee Fields sites in the Chattahoochee Valley, the 1 Ru 28 component is of minor importance.

A small amount of evidence indicates the presence of a Fort Walton-Lamar occupation similar in content to the Rood's focus (Caldwell 1955). Other ceramics are vaguely similar to Weeden Island or earlier Woodland materials.

The most interesting component at the site is a Late Archaic-Early Woodland manifestation. The sparse cultural remains from the other occupations made separation of this early component fairly easy, as the same materials would not have been so readily classifiable as Archaic if lost in a large collection of later overburden. A few projectile points, and especially some of the uniface tools, may be the result of other separate occupations; however, most of the projectile points and other chipped stone implements, steatite and sandstone sherds, the atlatl weight, hammerstones, grinding stones, etc., can be considered to be the remains of either a single component or a small number of closely related late Archaic components. The fiber tempered pottery was a terminal addition to the late Archaic: its presence gives the site transitional Archaic-Woodland status. The presence of Adena Narrow Stemmed and Pickwick Expanded Barb projectile points supports the indication of the ceramics that a transitional Archaic-Woodland occupation existed at this site. (Adena Narrow Stemmed has been found in both Archaic and Woodland strata in the Tennessee Valley sites.) Although Pickwick Expanded Barb was a strong late Archaic type at the Little Bear Creek site (Ct 8) and the Flint River site (Ma 48), examples were also recovered from Woodland zones at these sites (Webb and DeJarnette 1948a and b).

A somewhat earlier and less intensive occupation is represented by a few crudely chipped quartzite rounded base points, the jasper (Fs 100-2) and quartz-

TABLE 23

1 Ru 28 Non-Ceramic Artifact Catalog

Item	Catalog Number
SURFACE COLLECTION	
Artifacts of European Origin	
Gun flint, British pistol flint	142-1
Gun flint, possibly Indian-made	
Green glass bottle fragments (3)	
Projectile Points	
Base of medium-sized, rounded stem, quartzite point (Old Quartz Style), P-1	101-1
Base of medium-sized, rounded stem, quartzite Adena point	101-2
Base of a small, stemmed, serrated point	101-3
Tip of a large point	
Chipped Stone Tools	
Uniface end scraper 5.1 cm. long, 3.1 cm. wide	112-2
Ground Stone Artifacts	
Steatite vessel fragments (3)	
LEVEL 1 (0"-6")	
Artifacts of European Origin	
Iron nails, modern (3)	
Jew's-harp	
Metal fragments (3)	
Buckshot	
Green glass bottle fragments (3)	
Chinaware fragments (10)	
Projectile Points	
Short, stemmed, serrated point, 4.0 cm. long, 2.5 cm. wide	151-1

TABLE 23 (Continued)

1 Ru 28 Non-Ceramic Artifact Catalog

Item	Catalog Number
Short, broad, stemmed point, 4.3 cm. long, 3.2 cm. wide	118-1
Stem fragment of a large point	125-1
Tip of medium-sized point	167-1
Tip of point	153-1
Tip of point	105-1
Tip of point	105-3
Chipped Stone Tools	
Biface scraper 7.4 cm. long, 4.5 cm. wide	121-1
Biface core scraper, quartzite, 7.0 cm. long, 4.8 cm. wide	127-1
Broken biface scraper	137-1
Broken biface scraper	133-5
Crude drill, 3.6 cm. long	163-2
Flake knife 7.2 cm. long, 3.2 cm. wide	133-2
Flake scraper 4.0 cm. long, 3.4 cm. wide	163-1
Ground Stone Artifacts	
Steatite vessel fragment	151-3
Metate fragment	133-1
LEVEL 2 (6"-12")	
Artifacts of European Origin	
Metal fragment	130-3
Green glass bottle fragments (3)	
Chinaware fragments (2)	
Projectile Points	
Short, stemmed point 4.4 cm. long, 2.6 cm. wide	109-2
Base of a medium-sized, stemmed point	150-3
Base of a medium-sized, stemmed point	150-4
Tip of a medium-sized point	148-1
Short, broad, basal-notched, stemmed point 3.5 cm. long, 3.0 cm. wide	119-1

TABLE 23 (Continued)

1 Ru 28 Non-Ceramic Artifact Catalog

Item	Catalog Number
Medium-sized Elora point, 6.0 cm. long, 4.4 cm. wide	123-1
Short, stemmed, serrated point 3.2 cm. long, 2.6 cm. wide	122-2
Tip of Mississippian Triangular or Madison point	145-4
Tip of a medium-sized quartzite point	158-2
Tip of a very large quartzite point	147-3
Quartzite point 6.0 cm. long, 3.5 cm. wide	107-1
Quartzite point 5.5 cm. long, 3.2 cm. wide	109-1
Chipped Stone Tools	
Biface core scraper, quartzite, 8.2 cm. long, 5.9 cm. wide	149-1
Broken core scraper, quartzite	145-1
Broken biface scraper	145-2
Core	158-1
Ground Stone Tools	
Flaked and ground limestone axe 11.6 cm. long, 8.5 cm. wide	157-1
Worked hematite	150-2
Steatite vessel fragments (2)	
Sandstone vessel fragments (3)	
LEVEL 3 (12"-18")	
Artifacts of European Origin	
Rolled brass cone	108-1
Projectile Points	
Medium-sized, narrow, stemmed, serrated point 5.7 cm. long, 3.1 cm. wide	155-1
Stem fragment of a medium-sized point	155-2
Midsection of a medium-sized, narrow point	159-1
Base of a medium-sized, rounded stem point	159-2
Medium-sized, stemmed Elora point, 5.8 cm. long, 4.5 cm. wide	110-1

TABLE 23 (Continued)

1 Ru 28 Non-Ceramic Artifact Catalog

Item	Catalog Number
Tip of a medium-sized point	115-1
Short, wide, stemmed point 4.1 cm. long, 3.6 cm. wide	115-2
Medium-sized, expanded stem point 6.7 cm. long, 3.1 cm. wide	105-2
Midsection, small quartzite point	168-1
Tip of small point	105-1
Tip of a medium-sized point	105-3
Tip of a medium-sized point	117-2
Savannah River point, reworked to scraper	117-1
Midsection of point	120-1
<u>Chipped Stone Tools</u>	
Stemmed scraper, reworked from Savannah River point, 4.1 cm. long, 2.8 cm. wide	117-1
Biface scraper, rectanguloid, 3.7 cm. long, 2.8 cm. wide	161-1
Biface quartzite core scraper, 7.2 cm. long, 5.1 cm. wide	131-1
Core	162-1
<u>Ground Stone Tools</u>	
Worked hematite	154-1
LEVEL 4 (18"-24")	
<u>Projectile Points</u>	
Pickwick Expanded Barb point, 6.2 cm. long, 4.0 cm. wide	114-1
Savannah River point, 5.1 cm. long, 3.2 cm. wide	103-1
Adena Narrow Stemmed point	113-1
Appalachian Stemmed point made of pink jasper	100-2
Stem fragment of a medium-sized point	111-1
<u>Chipped Stone Tools</u>	
Small, biface scraper, quartzite, 4.3 cm. long, 2.3 cm. wide	100-1
Flake scraper, 4.6 cm. long, 1.2 cm. wide	152-2
Flake scraper, 4.3 cm. long, 2.5 cm. wide	152-1

TABLE 23 (Continued)

1 Ru 28 Non-Ceramic Artifact Catalog

Item	Catalog Number
Ground Stone Tools	
Pitted hammerstone	126-1
Metate	135-1
LEVEL 5 (24"-30")	
Projectile Points	
Tip of a medium-sized point	147-3
Tip of a medium-sized point	106-1
Quartzite, Appalachian Stemmed point	147-2
Chipped Stone Tools	
Small biface scraper, quartzite, 6.1 cm. long, 2.9 cm. wide	147-1
Ground Stone Tools	
Steatite atlatl weight	156-1
ARTIFACTS FOUND WITHIN THE UNDERLYING CLAY LAYER	
Projectile Points	
Depth: 30-36 inches	
Stemmed point, medium-sized, broken	132-1
Tip of medium-sized point	153-1
Depth: 3.31 feet	
Maples Broad Stemmed point, 9.3 cm. long, 5.1 cm. wide	124-1
Depth: 4.48 feet	
Medium to large-sized, stemmed point, 7.0 cm. long, 4.2 cm. wide	164-1
FEATURE 3	
Projectile Point	
Elora point, 5.1 cm. long, 3.2 cm. wide	

TABLE 23 (Continued)

1 Ru 28 Non-Ceramic Artifact Catalog

Item	Catalog Number
DEBRIS	
Projectile Points	
Basal section of medium-sized, stemmed point	104-1
Savannah River point, basal section	116-1
Chipped Stone Tools	
Biface core scraper, quartzite	141-1
Ground Stone Artifacts	
Steatite vessel fragments (3)	

ite (Fs 147-2) Appalachian Stemmed points, the quartzite Morrow Mountain Round Base point (Fs 101-1), and other tools of quartzite and flint. These, with the Maples Broad Stemmed and the undifferentiated, large, straight-stemmed point found in the deepest levels at the site represent the earliest component or components of the pre-ceramic occupation.

1 Ru 61, The Abercrombie Site

The Abercrombie site is located 6 miles south of Phenix City, Alabama, on the property of Mrs. T. K. Kendricks. The site, situated on the second terrace of the Chattahoochee River, consists of an eroded mound and adjacent village area. The area of the site is now a large pecan grove.

The site has received considerable attention, and is one of the better known archaeological remains in the Chattahoochee Valley. Clarence B. Moore visited the site in 1907. Peter A. Brannon conducted excavations there only slightly later. Wesley R. Hurt tested the site during his 1947 survey (when he assigned the numbers 1 Ru 10 to the village area and 1 Ru 1 to the mound) and the results of his work are incorporated here. Charles H. Fairbanks examined a surface collection from the site and his conclusions, published in *Early Georgia* (1955), are also examined here. More recently, the Columbus Museum of Arts and Crafts conducted extensive excavations, under the direction of Joseph Mahan, on the east side of the mound and along the continually eroding riverbank west of the mound. David Chase of the Fort Benning Infantry Museum has continued excavation in erosion-threatened areas adjacent to the river. The locations of his excavations are marked on the site map.

Previous work at the site had revealed the presence of two principal components: (1) a Historic Creek occupation evidenced by European trade goods in burials, a small amount of brushed pottery, and the presence of an incised ware resembling Ocmulgee Fields Incised; and (2) a Lamar component evidenced by the mound and a number of Lamar Complicated Stamped sherds.

The site is fairly large and extends over an area of about 100 yards in each direction. At the time of Clarence B. Moore's visit in 1907, the mound was estimated to be 14 feet high and 85 by 90 feet in diameter (Moore 1907: 149-150). During the years when Peter Brannon was excavating the site, the mound was 75 feet in diameter and 15 feet high (Brannon 1909: 189). Since that time, continual plowing has reduced the mound to about 8 feet in height. Moore describes the mound as having been domiciliary in nature, and his excavations did not reveal any artifacts. Hurt found a ridge of dirt about 80 feet from the center of the mound on the north side extending some 200 feet east and west. This ridge, as shown in Hurt's map of the site (Figure 57) ran tangent to the mound. In 1947, test pits made in this ridge suggested that it may have been a refuse mound covering a long row of house foundations on the west side of the mound. The remains of this ridge, shown in the contour map of the site, are not very prominent today.

PREVIOUS EXCAVATIONS

From August 16 to August 26, 1947, Hurt and a crew of Boy Scouts from Phenix City, Alabama, and Columbus, Georgia, excavated seven test pits in the area surrounding the mound and, in addition, cut a small trench in the west side of the mound. The major stratigraphic trenches of Hurt's excavations are shown in Figure 57. Below is a description of the features encountered in these test pits (Hurt Ms 1947).

Test Pits 1 and 2. These two pits actually formed one large test trench that was located 18 feet east of the mound. Test Pit 1 was excavated first and Test Pit 2 was added to the north. The length of these two pits was 165 inches north-south and 72 inches east-west. They were sunk to a depth of 30 inches—well into the sterile, underlying ground. The top 6 inches included the plowed zone of light sand mixed with clay. Sherds were scarce in this level. From 6 inches to 1.5 feet was a band of dark humus, containing many sherds and other refuse. In the northwest corner of Pit 2 a turtle shell was found 17 inches below the surface.

Test Pit 3. This pit was located 90 feet southwest of the mound at the west end of the long refuse heap. It was 8 feet long, 6 feet wide and 2.5 feet deep. The top 6 to 8 inches was the plowed zone, which contained a few sherds. The next 6 inches was a dark humus layer containing sherds and other refuse. Underlying this for several inches was a layer of very dark humus mixed with charcoal. Beneath the charcoal layer were four post molds, which extended into the underlying sterile clay level.

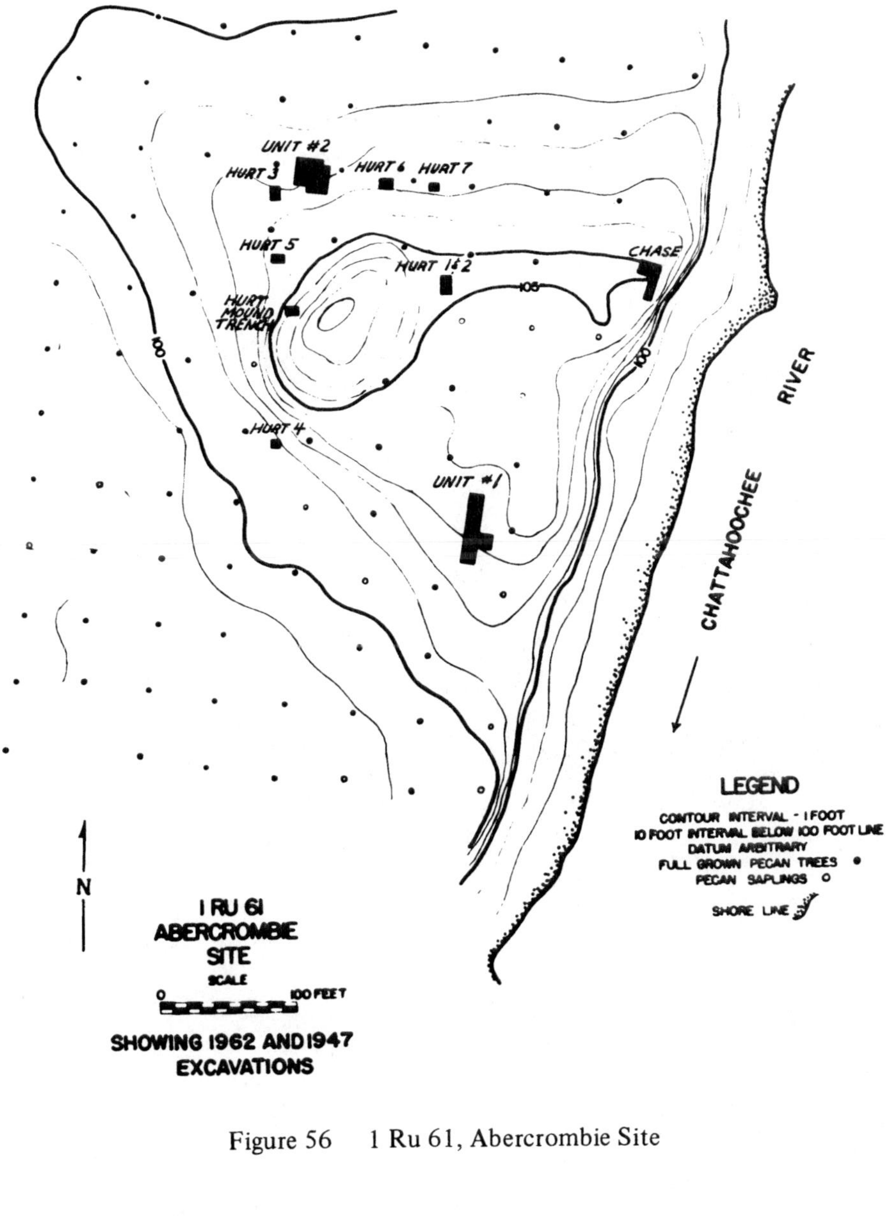

Figure 56 1 Ru 61, Abercrombie Site

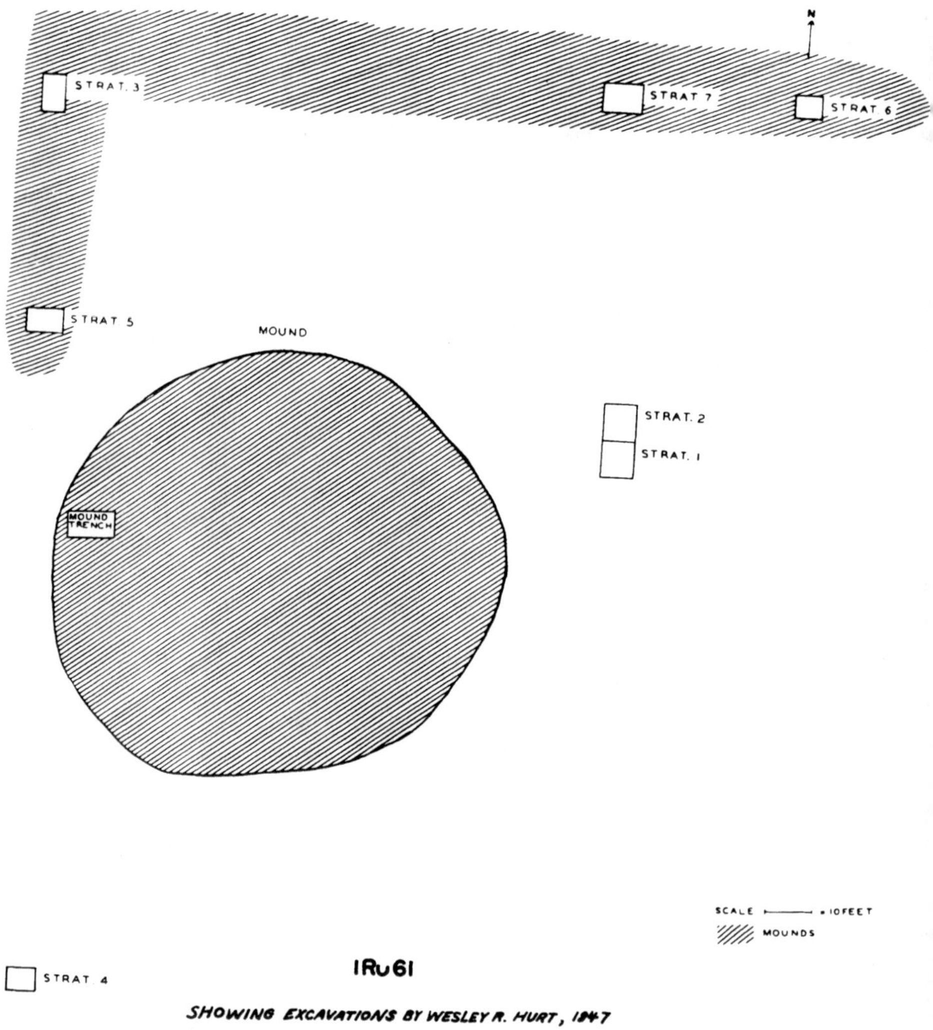

Figure 57 Hurt's 1947 Excavation

Test Pit 4. This pit was sunk 108 feet southwest of the mound. It was 6 feet by 5 feet by 3 feet deep. The top 6 to 8 inches of dark, plowed dirt contained several sherds. Underneath this lay a dark layer containing many sherds and other refuse items. The lower occupation level surmounted a band of sterile, yellow clay.

Test Pit 5. This pit was located 40 feet southwest of the mound and 42 feet south of Pit 4 on a line between Pits 3 and 4. It measured 8 feet long, 6 feet wide, and 3 feet deep. The top 6 to 8 inches was a plowed zone containing a few sherds. Underneath lay a dark, humus stratum, about a foot deep, which included refuse. In the center of the area, cut through this refuse zone, a cache pit a foot in diameter and a foot deep was encountered. It had been filled with sand, and a post hole had been dug in the fill at a later date. A layer of sherds extended through the refuse zone from the left edge of the cache pit. Underneath the refuse zone was a sterile yellow-clay band. A conoidal copper dangler was found in the plowed zone.

Test Pit 6. This pit was excavated into the east end of the refuse ridge. It lay about 225 feet east of Pit 3 and measured 6 feet by 5 feet by 42 inches deep. The top 6 to 8 inches was the plowed zone containing a few sherds and a rolled brass cone. Underneath the plowed zone was a foot of dark humus, rich in sherds, bone, and other refuse. The remains of three crushed Lake Jackson Plain bowls were found 6 inches deep in this humus. In the northeast corner and southwest corner of the pit were post molds, which extended to 6 inches below the plow zone. At the base of the refuse level, in the northeast corner of the pit was a shallow, rectangular fire pit that had been cut into the underlying sterile clay. This pit measured 2 feet across and 6 inches deep. This pit evidently was excavated in a house area.

Test Pit 7. This pit was 100 feet west of Pit 6 on the same long refuse ridge. It measured 8 feet long, 6 feet wide, and 3 feet deep. Stratigraphy had been disturbed in this area by burials and reburials. The top 6 inches, the plowed zone, contained a few sherds. At a depth of 4 inches, the remains of a burial (Burial 4) were encountered. The bones in this pit extended to a depth of 2 feet, where the sterile, hard clay began. Further descriptions of the contents of this pit are included in the burial section of this report.

Mound Trench. A trench 10 feet long and 5 feet wide was cut from the west edge of the mound toward the center. It had been excavated to the depth of only a foot when it became necessary for Hurt to abandon his project at Abercrombie site. This trench revealed that a large amount of refuse had been utilized in the mound construction, for sherds, bone, and broken artifacts were scattered randomly throughout the fill.

Mound Shell Deposit. On the surface, immediately southwest of the center of the mound, lay a deposit of shell. This deposit had been exposed the preceding spring when a drag had been run to level the mound. A pit approximately 40 inches square and 10 inches deep was sunk in this area. Forty freshwater mussel shells, sherds, bone, turtle shells, and antler fragments were uncovered. This shell deposit did not look like a ceremonial deposit, but rather resembled a place where kitchen refuse had been thrown on the mound in the process of building.

Records of other investigators of the site are not discussed at length here. David Chase has not yet [1963] completed his studies (the trench to the east of the mound next to the river was excavated under his direction), and Joseph Mahan has not yet published the account of his work at the site.

1962 EXCAVATIONS

Two trenches were excavated in the summer of 1962 by the University of Alabama field crew under Edward B. Kurjack. The first unit, 10 feet wide and 50 feet long, was located 200 feet southeast of the mound. The unit was enlarged in one area to permit the excavation of Burial 6. The second, measuring almost 25 feet square, was located approximately 25 feet north of the mound in the area of the midden ridge described by Hurt. The first trench yielded cultural material to a depth of 2 feet; the second unit was excavated to a depth of only 8 inches. Burials 6, 7, and 8 were uncovered in Unit 1; a number of post molds were present in Unit 2.

Structure Types

Post molds were uncovered by Hurt in his Test Pits 3, 5, and 6, and a number of other posts were encountered in Unit 2. No complete houses have yet been excavated, although numerous post molds and daub layers are being found in the area under excavation by Chase. The midden is particularly deep in that part of the site, and several daub layers with related post molds have been found overlying each other. Chase hopes that he will be able to describe the floor plan of a single house before his work at the site is completed.

The limited information available indicates that rectangular houses with relatively shallow, central fire pits and wall posts varying from 4 to 10 inches in diameter may have been present.

Hurt speculated that the midden ridge running perpendicular to the river was the remains of a row of houses, but only remnants of this ridge (shown on the contour map of the site) were present by 1962, so his hypothesis could not be further tested. Excavation of Unit 2 in the area of this ridge revealed the numerous post molds.

BURIALS FROM PREVIOUS EXCAVATIONS

C. B. Moore encountered a number of burials at Abercrombie Village in 1907:

> In the level ground around the mound is a cemetery in which we found, from one to two feet in depth, skeletons, some loosely flexed, some lying at full length on the back. There were also aboriginal disturbances where the bones of a skeleton had been disarranged by a burial made at a later period.
>
> The individuals whose skeletons were found by us evidently had experienced the advantage or disadvantage, of contact with Europeans, as many of the artifacts buried with them clearly proved.

Wesley R. Hurt included the following account of Moore's excavation at the Abercrombie site in his 1947 report:

> One skeleton had glass beads at the neck, and a bit of sheet-brass and a lump of hematite nearby; a steel or iron blade of a large knife on the body; a broad chisel of iron or of steel on the pelvis; and an iron or steel axe a little to one side of the body.
>
> Another skeleton at full length on the back had shell beads around the neck, which probably, when strung, supported a disc of sheet-copper, 4 inches in diameter, having a central hole, and two perforations for suspension near the rim which lay nearby. At an ankle and a wrist of the skeleton were a few shell beads. In the clay thrown out from the grave were two triangular sheet-brass pendants, and an annular ornament of the same material, 2 inches in diameter, and having an intervening space where the metal had broken between two perforations for suspension.
>
> Another skeleton had a large number of small sea-shells *(Marginella apicina)* pierced for suspension as beads.
>
> Above the skull of a skeleton was a small slab of stone, rounded to some extent, and four piercing implements of bone, three of which had the articular processes remaining. Below the skull was a triangular slab of stone and a large mussel-shell. Near the skeleton was an earthenware pipe of ordinary form, undecorated and of most inferior ware.
>
> Realizing that we had to do with burials of comparatively late Indians, we did not push the investigation of this place to the extent we would have done had the cemetery been of an earlier period. The mound we believe to antedate the post Columbian burial near it (Moore 1907: 450).

Hurt in his report of the Abercrombie site also comments on the work of Peter A. Brannon:

> Since 1909, Peter A. Brannon has excavated several burials at Abercrombie Village. He noted the occurrence of both flexed and extended burials and stated that '. . . with nearly every burial uncovered are found beads, gorgets, and other objects of adornment.' (Brannon 1909:190-191).

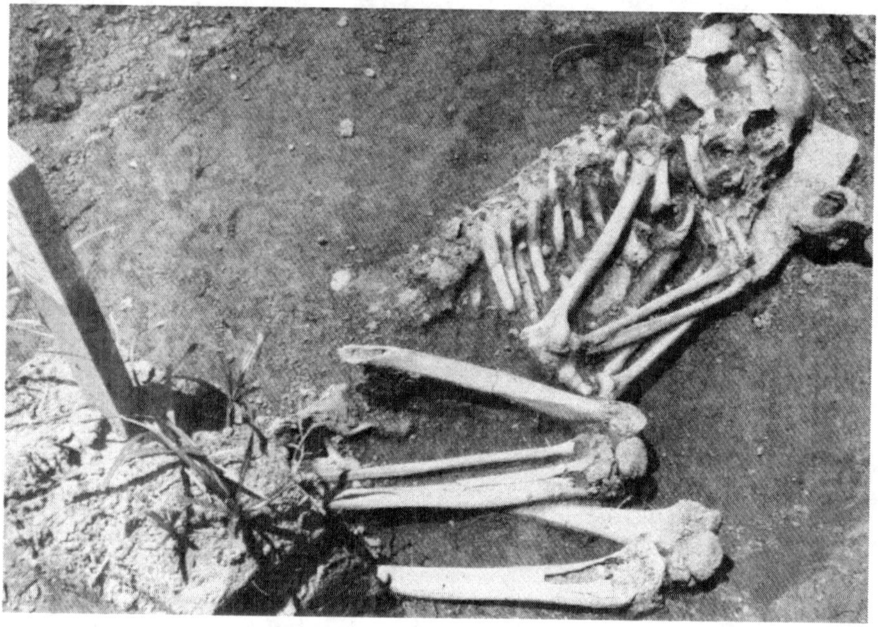

Figure 58 Burial 6 at 1 Ru 61 Showing Accompanying Iron Hoe

Figure 59 A View of the Excavations in Unit 1 at the Abercrombie Site, 1 Ru 61

During the excavations of 1947 at Abercrombie Village, four burials were found in Pit 5. Although all four were lying in a very restricted area, they seemed to be several distinct burials made over a period of time rather than a mass burial. Burial 4—its bones scattered and broken by plow action—lay on top of the other burials under only 4 inches of dirt. Its lowest bones were only 8 inches below the surface. Neither age nor sex could be determined.

Burial 1 lay below Burial 4 and was fairly complete, although the right side of the skull had been removed by deep plowing. The vertebrae were under 8 inches of fill. The skeleton lay in a twisted or semi-flexed position with the skull on the left side, the face to the east and top to the north. The torso rested on its back with the upper arm bones to the sides and the lower arm bones crossed over toward the east. The upper leg bones were thrust to the east so that the right tibia rested upon the left, while the lower leg bones were bent back towards the pelvis.

Miss Ruth Marzano, who studied this skeleton, stated that it was that of a male, 60 to 65 years of age. Fragments of the frontal and occipital bones showed marked frontovertico occipital deformation. Other pathological conditions included advanced arthritis (indicated by lipping on C-L vertebrae and complete ankylosis of two thoracic and two lumbar vertebrae) and dental caries and abscesses. According to Marzano, the latter condition is unusual among Southeastern Indians.

Burial 1 was accompanied by several artifacts. Between the upper arm and the ribs were two iron knives (one of which had an antler handle), one on each side. Between the left elbow and ribs was a deposit containing two ovoid and one flat polishing stone; one small, neatly made, chipped end scraper; two thin, finely made, triangular arrowheads; one drilled antler tip; and several cylindrical sections of antler that may have been "gaming pieces." Scattered over the skull were many small, blue, glass beads, while around the neck was a string of larger glass beads.

Burial 3 consisted only of a skull that lay to the west of the skull of Burial 1. It was, however, about 12 inches below the surface. The top of the skull was to the north, the face turned upwards. It had been smashed flat—probably by the weight of the overburden. The skull was that of a child 7 to 8 years of age who had caries. Beads of the type found with Burial 4 were also associated with this skull.

The skull of Burial 2 lay to the west of Burial 3. Its forehead—at one time 8 inches below the surface—had been removed by plowing. The top of the skull was oriented to the north with the face upwards and slightly to the south. The burial lay on its back with its legs twisted over in a semi-flexed position. Only the right side of the rib case was present, while the lower half of the left humerus was missing, along with the lower left arm and hand bones. The right humerus lay to the side of the ribs and the lower arm bones were crossed over the waist. The thigh bones were bent so that the right tibia rested on the left. They extended downward to the east at a 45-degree angle, while the lower leg bones were bent back adjacent on the south to the pelvic bones. Around the neck were three strings of the larger, blue, glass beads plus a smaller quantity of scattered beads of the smaller size.

Explanation for the missing left half of Burial 2 and the missing body of Burial 3 can be hypothesized. Burials 2 and 3 were probably interred simultaneously, side-by-side. Later, when a grave was dug for Burial 1, the bones now missing were probably thrown out. The grave for Burial 4 was probably dug on top of Burial 1. Moore noted in 1907 that several of the graves he excavated were intrusive into earlier burials (Moore 1907: 449).

Hurt believed the burials he excavated in 1947 were not directly associated with Abercrombie Village but belonged to the Historic Creek who occupied the area in later times. He based his hypothesis on the association of trade items with the burials, on the shallowness of the graves, and on their intrusion into the refuse from the old occupation of Abercrombie Village.

David Chase (personal communication) has uncovered several burials in the course of his work at the site. These burials differed, as a group, from those described above in four ways: (1) most were encountered at a greater depth; (2) no grave goods were present; (3) no European trade material was found in association; and (4) most of the burials were fully extended. Evidently most of these burials were related to the Lamar component at the site.

1962 BURIALS

Three burials were found in Unit 1 during the summer of 1962. Burial 6, shown in Figure 58, seemed to be the remains of a mature female. The burial, interred in a partly flexed position, was accompanied by an iron hoe. Burial 7 was found in an oval pit 2.2 feet long and 1.5 feet wide. A decayed longbone and several tooth caps were all that remained of the burial, but their placement suggested a partly flexed position. This burial was interred comparatively deeply—1.9 feet below the surface. With it was a narrow, ground, greenstone celt. Burial 8, that of a partly flexed child, was found 6 inches below the surface. This badly disturbed burial contained no artifacts.

POTTERY STUDY

The ceramic analysis of the Abercrombie site is contained in two studies, one based on the results of Hurt's excavation in 1947 and the other on the 1962 excavations.

Hurt's analysis appears in Table 25. (For a description of these types see Chapter 3 of Part I.)

Figure 60 1 Ru 61, Decorated Pottery Types

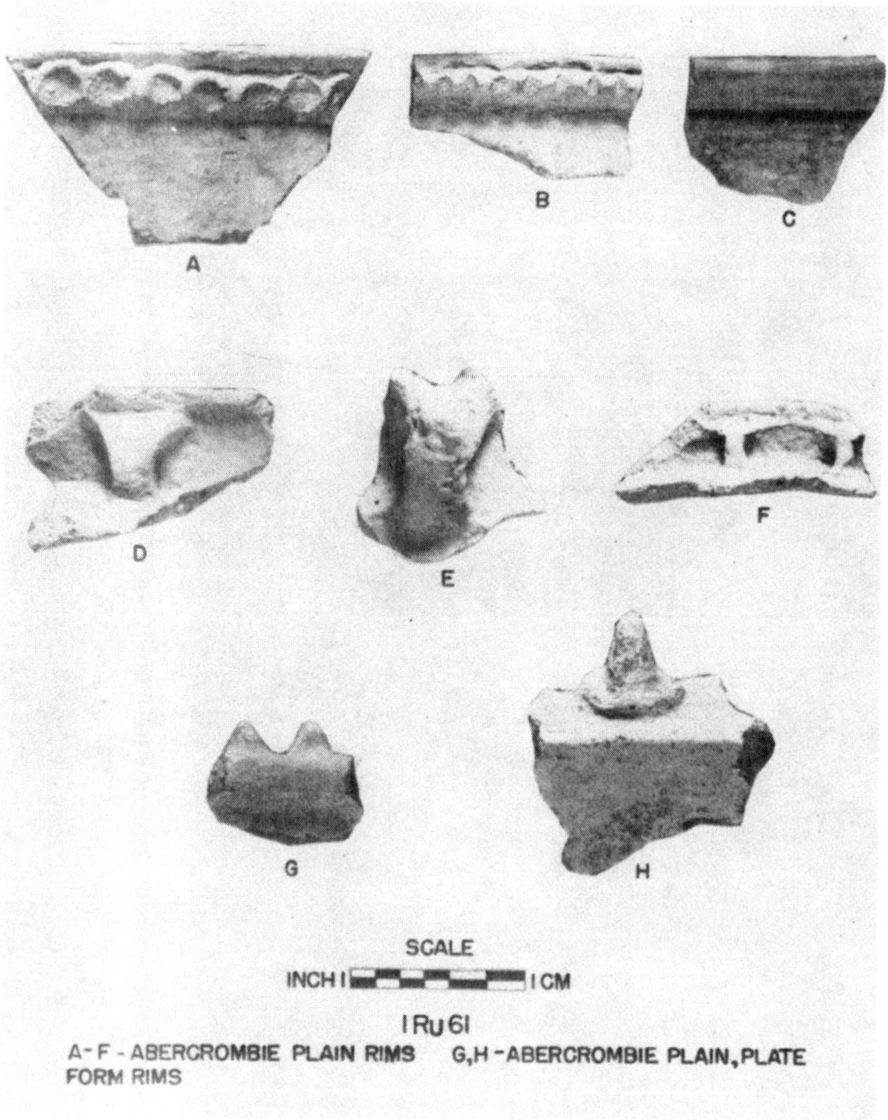

Figure 61 1 Ru 61, Abercrombie Rim Sherds

TABLE 24

1 Ru 61 (Abercrombie Site) Unit 2 Ceramics (1962)

	Number	Percent
Averett Plain	1	.129
Lamar Complicated Stamped	99	12.791
Lamar Bold Incised	4	.517
Lamar Plain	202	26.098
Fort Walton Incised	5	.646
Abercrombie Incised, Plate Form	3	.388
Abercrombie Incised, Casuela Form	69	8.915
Abercrombie Brushed	2	.258
Abercrombie Plain	210	27.132
Walnut Roughened	16	2.068
Red Filmed	4	.517
Black Filmed	41	5.297
Coarse Shell Plain	99	12.791
Fine Shell Plain	1	.129
Brushed Variant	3	.388
Miscellaneous Incised	3	.388
Miscellaneous Check Stamped	3	.388
Miscellaneous	10	1.292
Totals	775	100.000

NON-CERAMIC ABORIGINAL ARTIFACTS FROM HURT'S 1947 REPORT

Atlatl Weight. Fragment of a semi-lunar type of bannerstone, highly polished. 25 mm. long, 21 mm. wide, and 16 mm. thick. The association with Abercrombie Village was probably accidental.

Antler Tines. Four examples. Cut antler tines with points smoothed for use as flaking tools.

Awl, Bone. Small awl fragment made of a bird bone, broken at base, polished and fire-blackened. Length, 135 mm. Found in mound fill.

Axe, Stone. Peter A. Brannon found a three-quarter grooved axe at this site in 1906 (Brannon 1909: 192, Figure 41 *f*).

Bars, Stone. Two examples of cylindrical bars of unknown use. Both have rough surfaces. One is 29 mm. long, 7 mm. in diameter. The other is 47 mm. long and 6 mm. in diameter. Both were found in Test Pit 7.

Beads, Shell. Brannon illustrates several small pierced shells used for beads. They were of the shell *Marginella apicina* (Brannon 1909: 191, Figure 40).

Beads, Stone. According to Brannon, in the collection of Dr. H. H. Wheepley of St. Louis are many white-and-black-striped, stone beads from Abercrombie Village, some of them measuring nearly a half inch in diameter. More than a hundred came from one burial (Brannon 1909: 192-193).

Celt. Petaloid celt made of a dark-colored quartz. Only poll end found in 1947.

Gorgets. Brannon notes the occurrence of both shell and stone gorgets. He illustrates a gorget with three distinct pairs of broken-out suspension holes. In addition, he has found both pierced triangular and oval gorgets (Brannon 1909: 191-194, Figures 40, 43).

Hammerstones. Made from pebbles with pitted ends and polished faces. A single example found with Burial 2.

Hoe, Stone. Brannon illustrates a very fine stone hoe from Abercrombie Village. It was highly polished and showed no evidence of use. A hole was drilled through the center. This hoe accompanied a flexed burial in the cemetery and was associated with five thousand shell, bone, stone, and glass beads (Brannon 1909: 193-194, Figure 42).

In 1947, the fragment of a rectanguloid hoe made of hornblende schist was found. The surfaces were smoothed. It was 75 mm. long, 75 mm. wide, and 7 mm. thick.

Paint Pots. According to Brannon, a number of paint pots have been excavated at the Abercrombie site, although he doesn't state what type they are (Brannon 1909: 193). The writer observed in various private collections in the Chattahoochee Valley a large number of small oval stones that were evidently used as paint pots, although he found none at Abercrombie Village. These stones have a circular depression on one or both sides and are often worn all the way through. Since all these artifacts were surface finds, it was impossible to determine their provenience.

Pendants, Columella. Teardrop-shaped pendant of a shell columella. Smoothed surface. Broken. 40 mm. long, 14 mm. wide, and 9 mm. thick.

Pipe Bowl, Pottery. Fragment of a cylindrical pipe bowl or stem. Smooth surface. 30 mm. long, 34 mm. wide, and 8 mm. thick. Red color. From Test Pit 6, 6-inch to 1-foot level. In Test Pit 7 fragment of a similar bowl 39 mm. long, 30 mm. wide was found.

Pipe, Stone. Brannon described a stone pipe from this site as a "hammer pipe" of white quartz, highly polished inside the bowl and around the base. "The object is round, about 4 inches high, nicely grooved, and although perfectly hammer-shaped, it is at the same time clearly a pipe" (Brannon 1909: 194).

Polishing Stones. Small pebbles with highly polished faces. Two examples associated with Burial 2.

Projectile Points. Several varieties found. These include the six examples of the small triangular points (Hurt's type NB2), two examples of the very slender triangular point (Hurt's type NC), one example of a corner-notched point (Hurt's type SG).

Scraper, Shell. Fragment of a scraper made of a mussel shell. Sharp cutting edge and polished surface. 34 mm. long, 25 mm. wide, and 4 mm. thick.

Scraper, End. Small ovoid-shaped, flaked end scraper.

Sherds, Worked. Two examples found, one disc-shaped and the other disc-shaped with a hole through center. From Test Pit 4, 1-foot to 1-foot-6-inch level.

Spatula, Bone. Made of mammal cannon bone with one beveled end. Highly polished. 84 mm. long, 20 mm. wide, 10 mm. thick. Broken. From Test Pit 6, 6-inch to 1-foot level.

NON-CERAMIC ABORIGINAL ARTIFACTS 1962 EXCAVATION

Fifteen stone and bone artifacts of aboriginal manufacture were found at 1 Ru 61. These included projectile points and bone and ground stone tools.

Projectile Points. Nine whole or fragmentary points were recovered. Two of these were of an earlier—possibly Archaic—component; five were Mississippian triangular types; two were fragments that appeared to have been unfinished. (See Figure 62.)

Bone Artifacts. Two bone beads and the tip of a bone awl or chipping tool were found.

Ground Stone Tools. One greenstone celt, one fragment of a polished axe or celt, and one crudely flaked, axe-like tool were present at the site.

TABLE 25

Hurt's Analysis of Ceramics From Abercrombie Village (1947)

	Mound Trench		Test Pits 1 and 2			
Type	0"-6"	6"-12"	0"-6"	6"-12"	12"-18"	18"-24"
Fort Walton Incised	1	1				
Abercrombie Incised			1			1
Chattahoochee Brushed			1			
Chattahoochee Brushed Sub-Type A						
Lamar Complicated Stamped	20	35	3	3	8	
Lake Jackson Plain	19	34	47	2	67	2
Carrabelle Incised						
Coweta Micaceous Utility						
Comer Cord Marked						
Ocmulgee Fields Incised	5		16	5	27	5
Ocmulgee Fields Incised Sub-Type A			2			
Ocmulgee Fields Incised Sub-Type B				1	1	
Ocmulgee Fields Incised Sub-Type C						
Abercrombie Black Filmed	2				17	
Abercrombie Shell Tempered	62	6	5	6	7	5
Total	109	76	75	17	127	13

Test Pit 3				Test Pit 4		
0"-6"	6"-12"	12"-18"	18"-24"	0"-6"	6"-12"	12"-18"
4	2				1	4
1	2			2	2	
					1	
1	2					
	2					
46	33	8	2	83	1	9
355	58	54	5	116	34	53
2	11			10		9
2						
50	10			32	7	17
				3	2	
1				1		
1				1		
	17			20		
100	67	4	1	146	3	23
563	204	66	8	414	51	115

TABLE 25 (Continued)

	Test Pit 4				Test Pit 5		
	18"-24"	24"-30"	30"-36"	36"-42"	0"-6"	6"-12"	12"-18"
Fort Walton Incised	1	3	1		3	1	
Abercrombie Incised			1				
Chattahoochee Brushed	3		1				
Chattahoochee Brushed Sub-Type A			1		1		
Lamar Complicated Stamped	10	4	20	14	62	52	4
Lake Jackson Plain	62	15	10	40	277	163	11
Carrabelle Incised							
Coweta Micaceous Utility						17	
Comer Cord Marked							
Ocmulgee Fields Incised	10	3	1		20	5	1
Ocmulgee Fields Incised Sub-Type A					1		
Ocmulgee Fields Incised Sub-Type B					1		
Ocmulgee Fields Incised Sub-Type C							
Abercrombie Black Filmed	8	3			6	10	
Abercrombie Shell Tempered		6	6	2	29	13	8
Total	94	34	41	56	400	261	24

Test Pit 6					Test Pit 7		
0"-6"	6"-12"	12"-18"	18"-24"	24"-30"	0"-6"	6"-12"	12"-18"
	3					1	5
					1		
		1					
	1	4					
					15	10	16
54	23	32	9	11	159	182	54
192	58	230	157	41			
	1						
					44	24	5
13	35	12		1			
		1					
	1						
					14	15	11
		7	3				
					109		
22	10	10	6				
281	132	297	175	53	342	232	91

Figure 62 1 Ru 61, Miscellaneous Aboriginal Artifacts

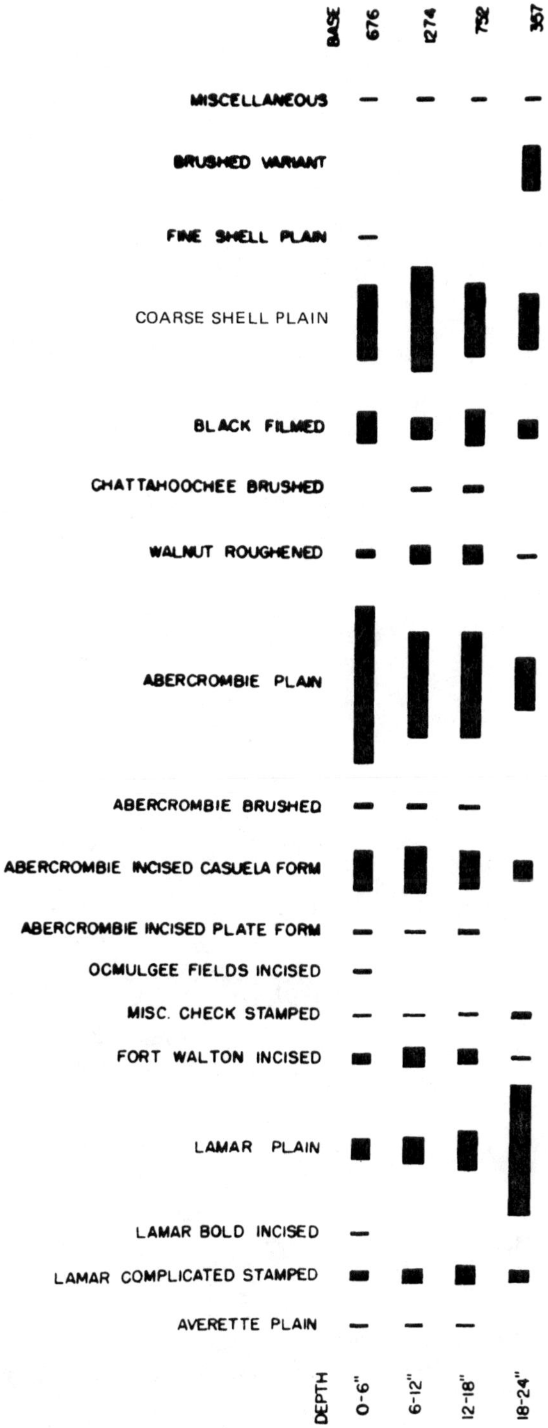

Figure 63 1 Ru 61, Abercrombie Site, 1962 Ceramic Depth Distribution Diagram

EUROPEAN TRADE MATERIAL

Fair quantities of European-made artifacts have been reported from the Abercrombie site, but only a single item of this category—an iron hoe found in burial association—was recovered during the 1962 excavations at the site.

Few trade objects have appeared on the surface or in the general diggings at Abercrombie, and almost all of these have been associated with burials. A list of the various items found at the site follows.

Claypipe Fragments. Hurt found one claypipe bowl fragment in Test Pit 1 at the 6-inch to 1-foot level.

Glass Beads. Hurt reported the presence of glass beads with Burials 1, 2, 3, and 4. He mentioned three types: (1) a single example of a Georgia Milk Oval bead (DeJarnette and Hansen 1960); (2) one Carnaline d'Aleppo Red on Green Barrel (DeJarnette and Hansen 1960); and (3) a large number of turquoise-green, Itchtucknee Blue beads.

Gun Flint. One well worn gun flint made from waxy-gray, British (?) flint was found by Hurt. It was of pistol-flint size.

Iron Tools. Several types of iron tools have been found in burial associations. An iron hoe, shown *in situ* in Figure 58, was recovered with Burial 6. Hurt reported an iron knife in association with Burial 1. Moore mentions an iron axe and chisel from the site.

Sheet Brass Ornaments. One rolled, sheet brass cone was found by Hurt in the first 6-inch layer of Test Pit 6. Moore mentions triangular pendants and a perforated, circular disc.

Although European goods were common in burial associations at the site, they were rare in the village refuse. This scarcity of European artifacts in the refuse may indicate that trade goods were scarce among the village Indians. The gun flint found by Hurt was the only indication of the use of firearms. Ornaments and iron hand tools were found in considerable quantity, while claypipe fragments, glass bottle fragments, and parts of firearms—all common at middle eighteenth century sites—were almost absent.

CONCLUSION

The aboriginal ceramics collected during the 1962 excavations at 1 Ru 61 can be placed generally in three cultural horizons: Lamar, Ocmulgee Fields, and Abercrombie. Lamar and Abercrombie pottery types are represented rather equally and make up the bulk of the sherd count; Ocmulgee Fields constitutes a minority. (See Figure 63.) The most abundant Lamar decorated type was Lamar Complicated Stamped. The most common incised type was the variant type

Abercrombie Incised in the plate form. Artifacts of European trade usually accompanying Ocmulgee Fields pottery types were rather sparse.

The 1962 excavations at the Abercrombie site (1 Ru 61) tend to support Hurt's suggestions (Part I) that the culture at Abercrombie Village may be coeval with or even older than the Lamar at Macon Plateau and that quite possibly Abercrombie Village was occupied during the time when the Creeks were first migrating across Alabama and into Georgia, rather than dating from the time when they moved back into Alabama after the Yamassee War in 1715. The predominance of Lamar Complicated Stamped and Abercrombie Incised in both Units 1 and 2, excavated in 1962, would suggest that rather than being older than Lamar, Abercrombie represents a blending of central Alabama traits with those of the Georgia Lamar.

1 Ru 66, The Patterson Site

The village areas in the vicinity of the Spanish Fort site were tested in three places. First, the area immediately adjacent to the fort, site 1 Ru 66, was tested with several pits by the Smithsonian Institution's crew. Then in the fall of 1961, a series of trenches and test pits was excavated 200 yards southeast of the fort. This excavation, also designated site 1 Ru 66, yielded a collection of Ocmulgee Fields pottery and a considerable amount of European trade material. Finally, during the summer of 1962, test pits were excavated in the second adjacent field, 600 yards west of the fort. Four test pits were dug over an area 300 feet across, but only a few plain, grit tempered sherds rewarded this effort.

The excavations at 1 Ru 66 consisted of seven test trenches and four test pits. The layout of these excavation units is noted in the map, Figure 66. Cultural material was taken from two 4-inch levels at the site.

1 Ru 66 was a relatively shallow site. The plow zone, containing the cultural material, measured 4 to 8 inches in depth; below the plow zone, hard, yellow clay was encountered.

Two features were found at the site. Feature 1, shown in Figure 67, was a basin-shaped refuse pit, approximately 2 feet deep and 5.5 feet in diameter. Feature 2 was a similar pit, more irregular in outline and slightly shallower. It measured 4 feet in maximum diameter and 1 foot in depth. Both features contained European trade materials: Feature 1, a large iron fragment and a gun flint; Feature 2, a claypipe stem fragment and two iron fragments. Large, mammal bone fragments, most of which appeared to be deer bone, were found in both pits.

POTTERY

Although pottery found in the excavation units at 1 Ru 66 included all of the types found within the fort, shell tempered pottery was less abundant in the village areas and considerable amounts of grit tempered plainware and Chattahoochee Brushed pottery were collected. Ocmulgee Fields Incised sherds found at the village area tended to contain more grit in their temper than Ocmulgee

Figure 64 Feature No. 1, a Basin-Shaped, Refuse Pit at 1 Ru 66

Figure 65 A View of the Excavations in Trench 1 at 1 Ru 66

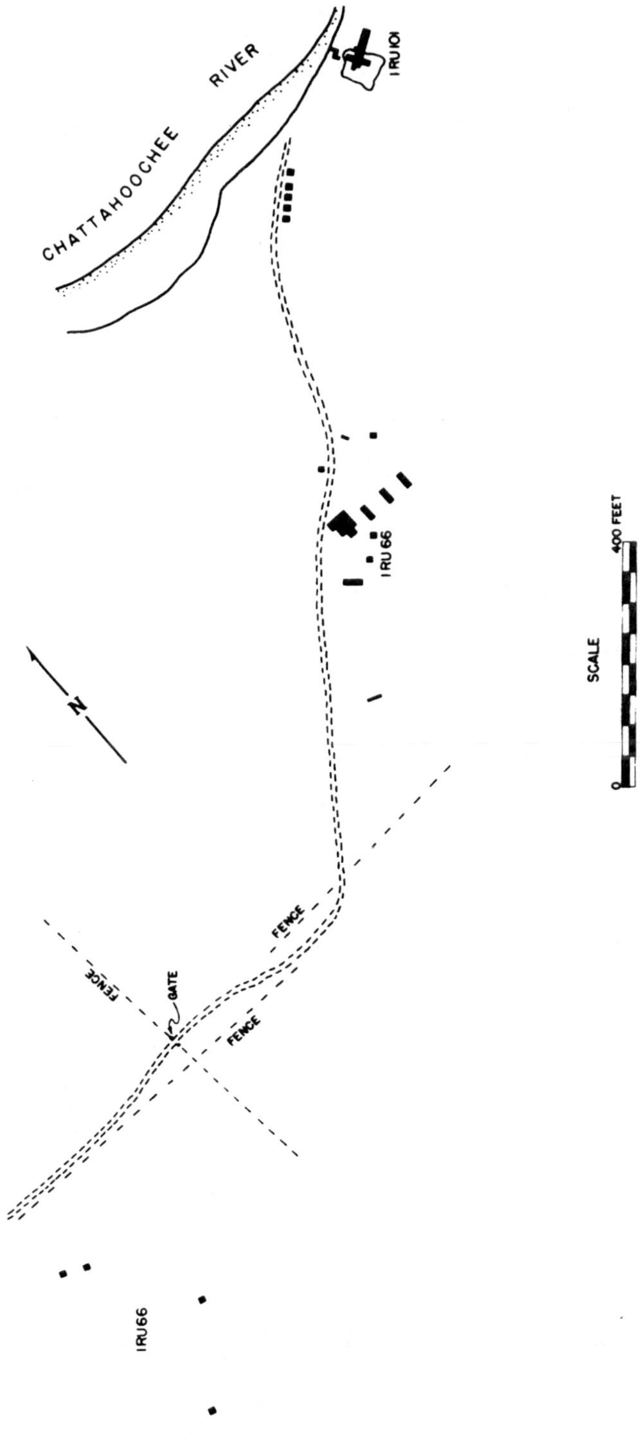

Figure 66 1 Ru 66, Excavations in the Area of the Spanish Fort Site

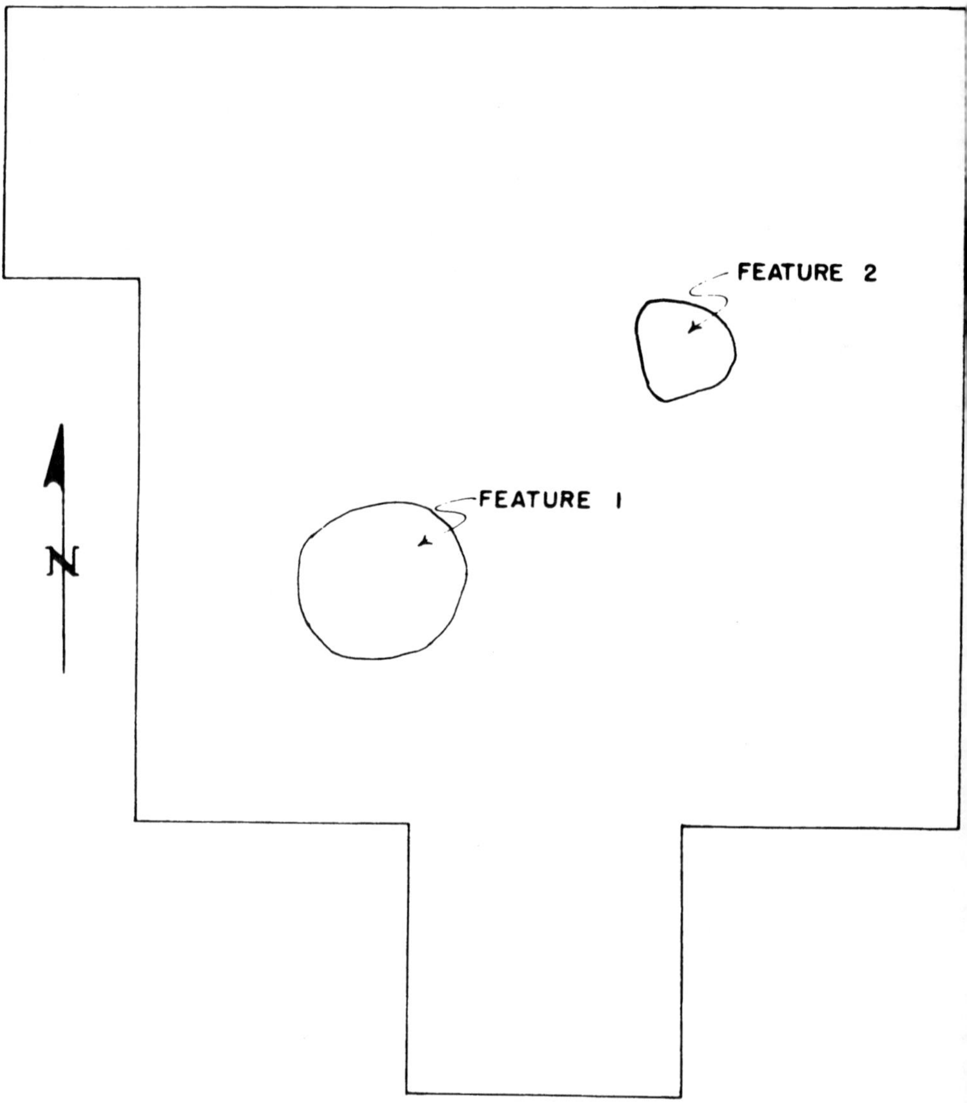

Figure 67 1 Ru 66, Unit 1

Fields Incised sherds at the fort.

No significant difference in depth distribution of pottery types was observed at the site. A frequency distribution of the various types is shown in Table 26.

NON-CERAMIC ABORIGINAL ARTIFACTS

Ground and Chipped Stone Artifacts. Several ground and chipped stone artifacts were found at 1 Ru 66. While some of these were related to the Historic period, the majority were not. Three fragments of a large, stemmed type of projectile point; one end scraper; a small knife or Willow Leaf Blade; one "T" drill; three small, unfinished blades or blanks; a fragment of a small, round-stemmed point; and a tip fragment of a small projectile point were recovered. The size and shape of the small projectile point fragments indicate that they were late types. The three large, stemmed projectile point fragments were of an earlier–possibly Archaic–component. Two fragments of celts or axes were found, but they were so incomplete that they could not be classified.

Metal Ornaments. Aboriginally manufactured metal ornaments consisted of one rolled sheet copper bead.

TABLE 26

Total Sherd Count, 1 Ru 66

	Number	Percent
Ocmulgee Fields Types		
Ocmulgee Fields Incised	570	7.314
Chattahoochee Brushed	505	6.480
Walnut Roughened	204	2.618
Kasita Red Filmed	47	.603
Ocmulgee Fields Plain, Coarse Tempered	4992	64.057
Ocmulgee Fields Plain, Fine Tempered	830	10.651
Plain Shell Tempered	593	7.609
Miscellaneous		
Lamar Complicated Stamped	37	.475
Eroded Complicated Stamped	12	.154
Stallings Plain	3	.038
TOTAL	7793	100%

TABLE 27

Pottery From Refuse Pits, 1 Ru 66

Type	Feature 1	Feature 2
	Number	Number
Ocmulgee Fields Incised	18	6
Chattahoochee Brushed	10	11
Walnut Roughened	6	
Kasita Red Filmed	3	
Lamar Complicated Stamped	3	
Plain Shell Tempered	29	1
Ocmulgee Fields Plain		
Coarse Tempered	153	28
Fine Tempered	36	18
Total	258	64

ARTIFACTS OF EUROPEAN MANUFACTURE

Many items of European manufacture were found at the site, several of them in association with Indian-made articles. The European objects included glass wine bottle fragments, claypipe fragments, gun flints and gun parts, a brass "scallop shell" ornament, various types of European ceramics, and iron tools (spikes, nails, knife blade, large staple).

Gun Parts. Among the gun parts at 1 Ru 66 were a fragment of a flintlock cock, two musket lock springs, a sear, and a fragment of a brass ornamental side plate. Four gun flints–two of Indian and two of European manufacture–were found. The European flints appeared to be made of dark gray English flint.

European Ceramics. Seventeen fragments of various types of European ceramics were found at 1 Ru 66. These included six fragments of English Creamware, five fragments of Blue Feather Edge, one fragment of Green Feather Edge, one fragment of Whieldon Tortoise Shell (1750-1770), one fragment of Annular Ware, and three fragments of German Gray Stoneware. Except for the Gray Stoneware and English Creamware, both of which were also manufactured during the first half of the eighteenth century, these types all date from approximately 1750 to 1800.

Claypipe Fragments. Eight kaolin pipe stems (all with bore diameter of 5/64 of an inch) and three bowl fragments were recovered at 1 Ru 66. One bowl frag-

Figure 68 1 Ru 66, Ocmulgee Fields Incised and Sand Tempered Plain Sherds

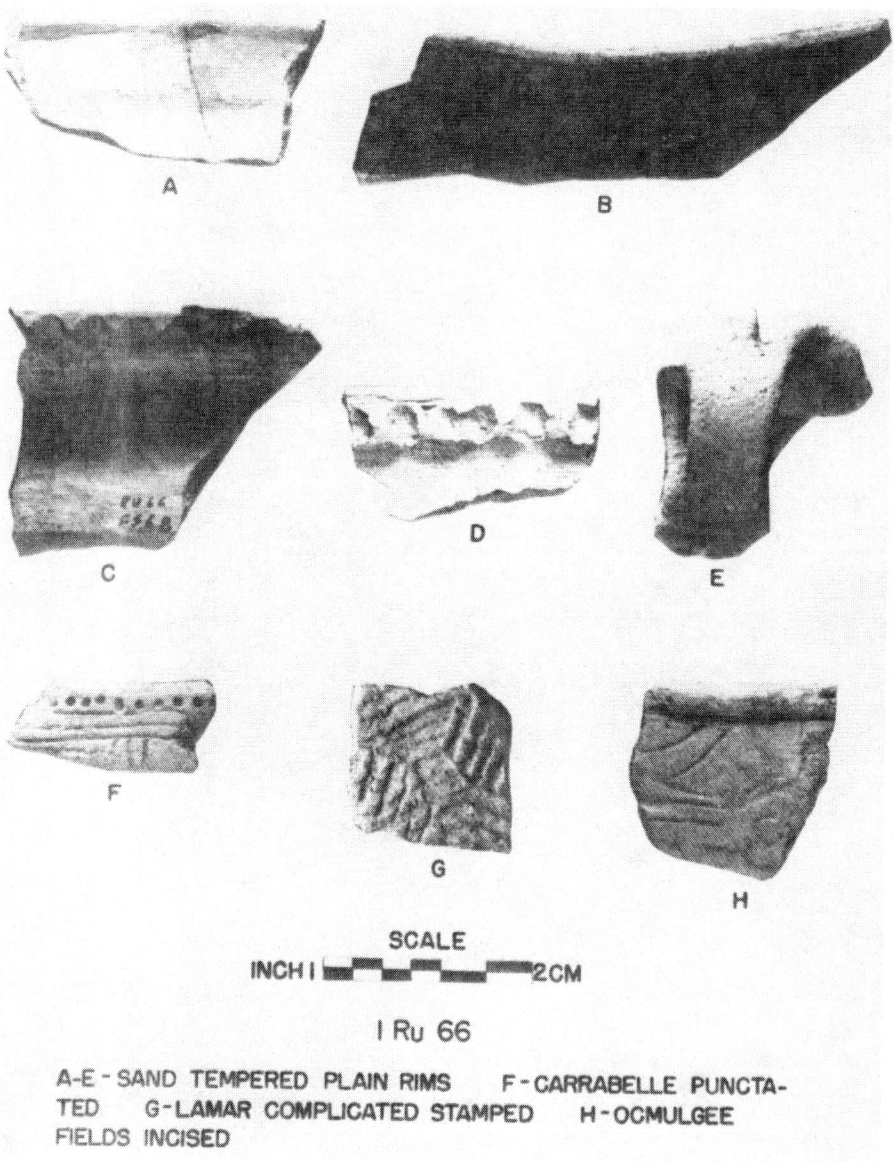

Figure 69 1 Ru 66, Miscellaneous Pottery Types

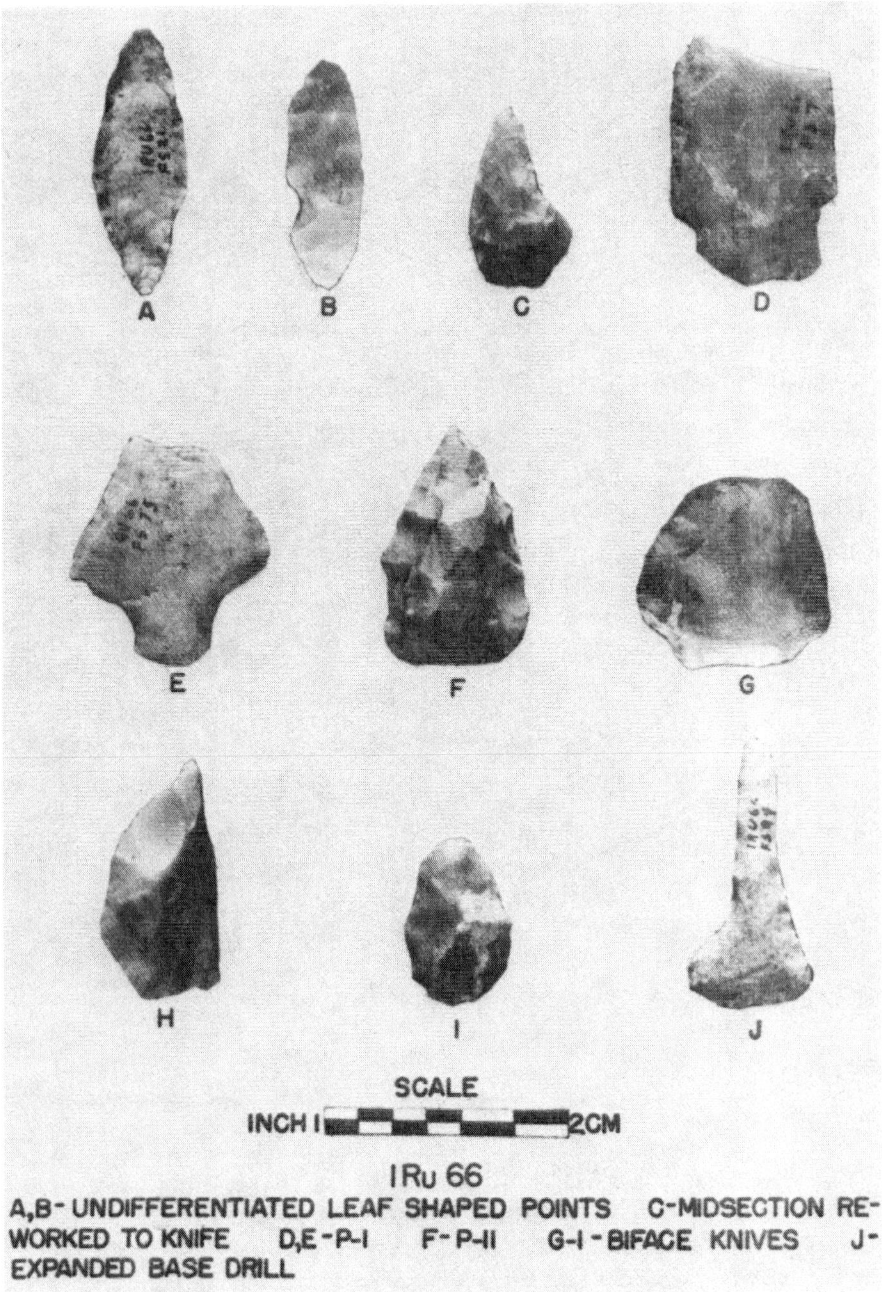

Figure 70 1 Ru 66, Flint Artifacts

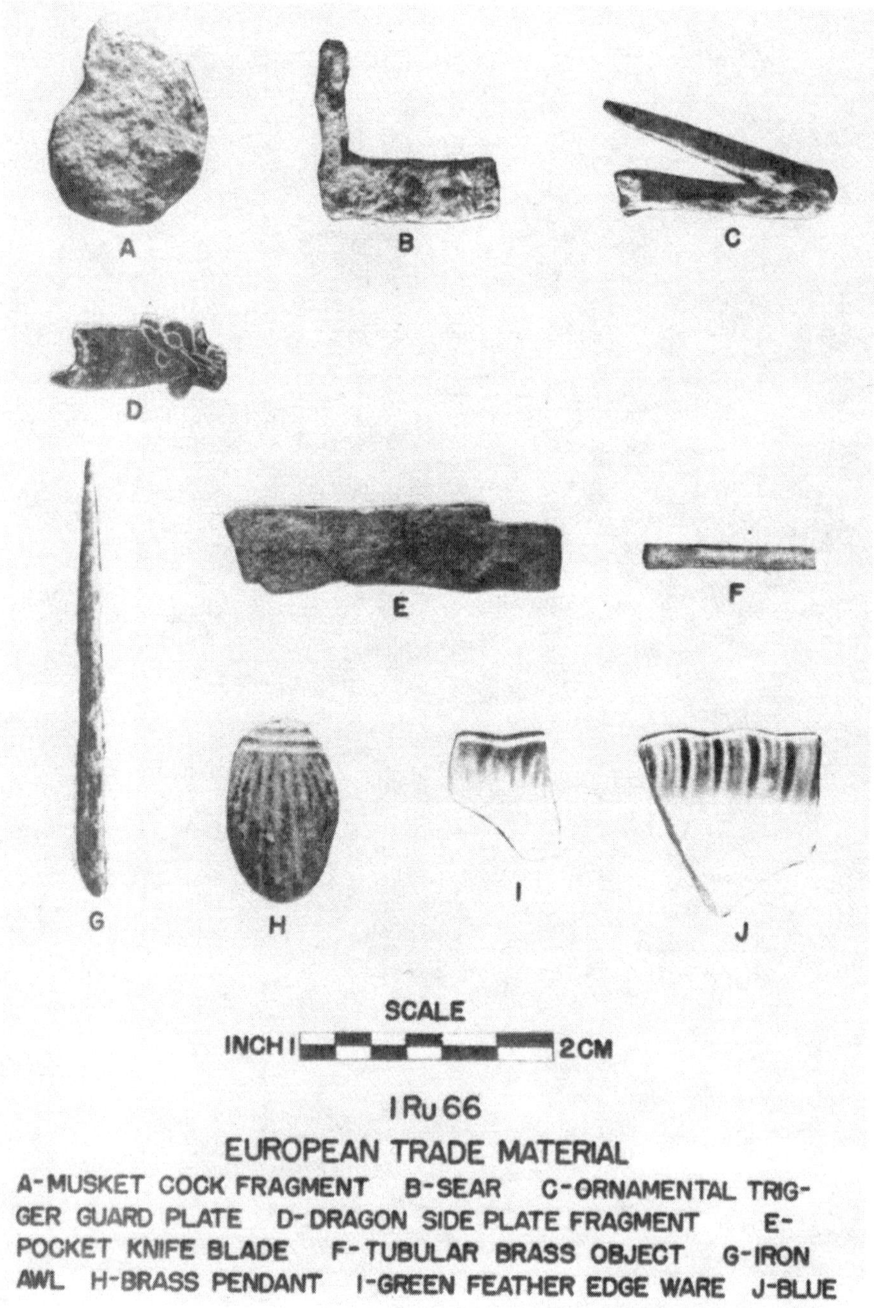

Figure 71 1 Ru 66, European Trade Material

ment bore the stamped letters "R.T.," initials of Robert Trippet, III and mark of his well-known Bristol pipe manufacturing concern, which was in operation as late as 1724 (Charles H. Fairbanks, personal communication). This type of pipe was also found at the Childersburg site (1 Ta 1).

CONCLUSIONS

The village areas outside the Spanish Fort site seem to show more evidence of later occupation than does the fort. The pottery type Chattahoochee Brushed, almost completely absent from the collections made within the fort, is present in quantity at 1 Ru 66. More European trade objects were collected in the village area, and all of them probably date from the middle and late eighteenth century. Aboriginal ceramics much like those found in the fort form a large part of the total collection. 1 Ru 66, therefore, probably represents an intermittent occupation lasting through the latter part of the seventeenth century into much of the eighteenth century.

1 Ru 69, The Cottonton Site

Site 1 Ru 69 is situated on the north bank of the Chattahoochee River in a bend located east of Cottonton, Alabama. Limited excavations at the site were conducted by a student crew from Florida State University during the summer of 1960. Exploration of 3,000 square feet of village area revealed a sparse concentration of Historic Creek pottery and European trade materials.

A formal grid system was employed at the site (the area and layout of the excavations are indicated in the contour map and floor plans–Figure 75), and cultural material was collected from two 6-inch levels. The site was relatively shallow, consisting of 2 to 6 inches of disturbed, culture-bearing soil covered by a 6-to-8-inch plow zone.

Sergeants William Phillips and Samuel Firenza of a mine warfare training unit at Fort Benning visited the site during the excavations. Using a mine detector, they located several metal objects at depths of 1 to 3 inches.

POTTERY

The base of a deep Chattahoochee Brushed pot (Figure 74) was restored from sherds found together at the site. (The top part of the vessel was carried away by the plow.) The restored portion of the vessel is approximately 7 inches in diameter and 5 inches deep.

All of the 1,624 sherds collected at the site were easily classified into the Ocmulgee Fields types common to the area–Ocmulgee Fields Incised, Chattahoochee Brushed, Kasita Red Filmed–as well as fine and coarse Ocmulgee Fields plainware. The types seemed identical to the Historic component pottery at 1 Br 34.

A depth distribution of the sherds is shown in Figure 77. The smaller proportion of brushed sherds in the deeper, 6-to-12-inch level may indicate increased popularity of the type during the more recent occupation of the site.

Figure 72 Site 1 Ru 69

Figure 73 Excavations at 1 Ru 69

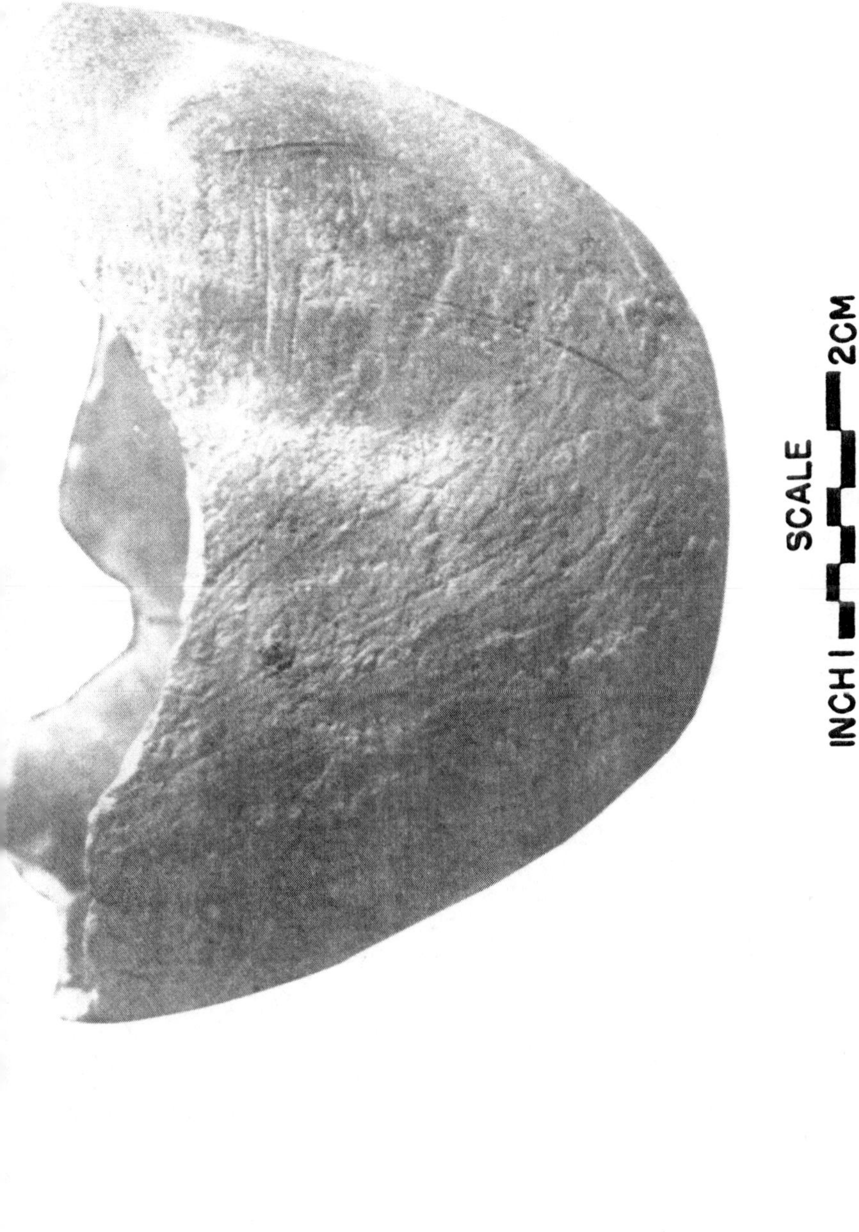

Figure 74 1 Ru 69 Chattahoochee Brushed Vessel

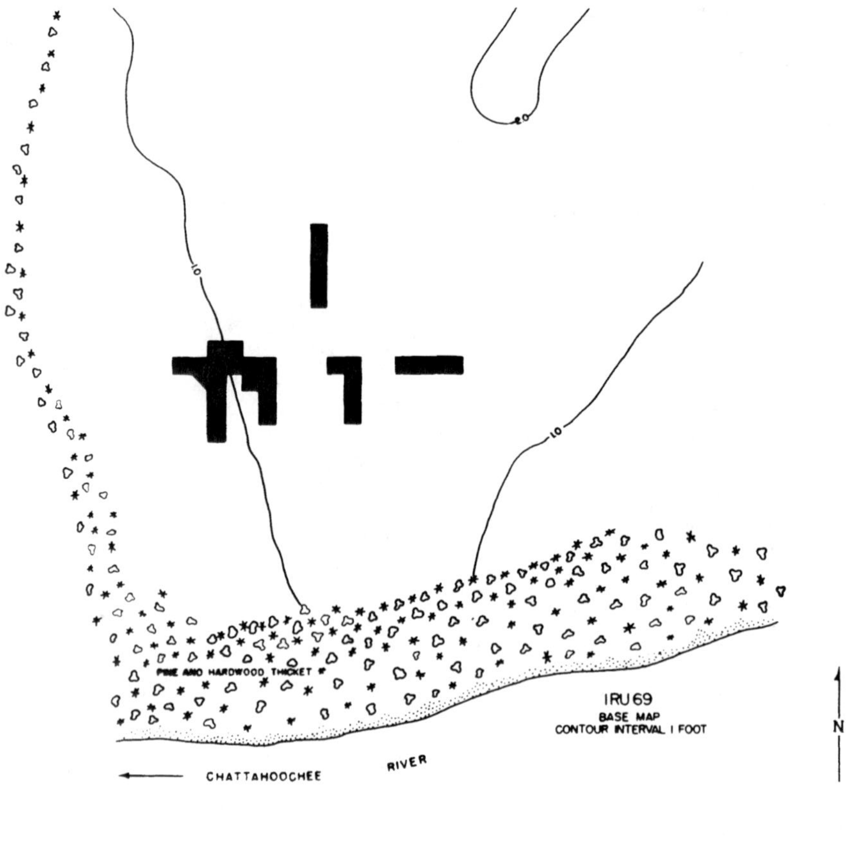

Figure 75 1 Ru 69 Base Map

Figure 76 1 Ru 69, Miscellaneous Ceramic and Chipped Stone Artifacts

NON-CERAMIC ABORIGINAL ARTIFACTS

The two large, stemmed projectile points illustrated in Figure 76 seem to represent a late Archaic component at the site. One other fragment, the distal end of a small thin triangular point, was found. A steatite sherd, a steatite pipe fragment, and a fragment of a ground celt were the only ground stone artifacts recovered.

EUROPEAN TRADE MATERIAL

A great deal of European trade material was encountered at 1 Ru 69. Much of it was found on or near the surface, mixed with artifacts of aboriginal manufacture. Many of the metal objects near the surface were located with the mine detector generously lent by United States Army authorities at Fort Benning, Georgia.

Among the artifacts recovered were iron tools, gun parts and gun flints, glass wine bottle fragments, brass objects, European ceramics, and claypipe fragments.

Iron Objects, Tools. Four large nails, an axe handle wedge, and what appeared to be a portion of a trunk hinge were found at the site. The nails were hand wrought.

Gun Parts and Gun Flints. Gun parts found at 1 Ru 69 included one lock and frizzen, one brass trigger guard fragment, one brass butt plate fragment, one ornamental side plate fragment, one iron trigger, and eleven gun flints of various sizes.

The style of the parts resembles that of English trade muskets; the brass trigger guard fragment and side plate fragment are similar to the examples in the collec-

TABLE 28

Pottery Count, 1 Ru 69

Type		Number	Percent
Ocmulgee Fields Incised		17	1.047
Chattahoochee Brushed		568	34.975
Kasita Red Filmed		47	2.894
Ocmulgee Fields Plain			
Coarse Tempered		384	23.645
Fine Tempered		607	37.377
Stallings Plain		1	.062
	Total	1,624	100.000

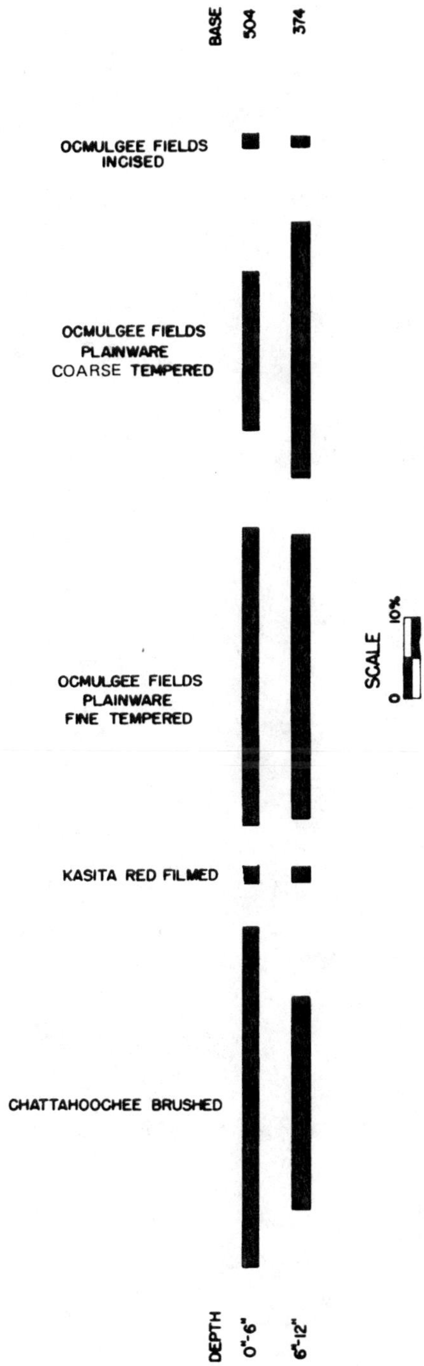

Figure 77 1 Ru 69, Depth Distribution of Pottery Types

tion of trade muskets in the "Powder Magazine" in Colonial Williamsburg. The relatively inferior quality of these gun parts also indicates that they may have come from trade weapons.

Some of the gun flints were crudely made; this may have been the result of Indian manufacture.

Glass Wine Bottle Fragments. Fifteen fragments of wine bottles were recovered from the surface and first level of occupation at the site. These all appeared to be fragments of a green glass variety known to have been used during the eighteenth century.

Brass Objects. A brass rivet and four pieces of sheet brass or copper were collected. The sheet metal fragments may have been salvaged from kettles or pots for reuse in other items.

European Ceramics. One small fragment of a German Gray Stoneware ceramic jug with salt and cobalt blue glaze was found at 1 Ru 69.

Tobacco Pipes (European and Aboriginal). Only one stem fragment of white kaolin pipe was found at 1 Ru 69. Three aboriginal pipe fragments were encountered. One of these, of pottery, indicated a square stem having small punctates; another fragment appeared to be from a round bowl pottery pipe; and the remaining fragment was of steatite, apparently from the pipe bowl.

CONCLUSIONS

The majority of the cultural materials at the site give evidence of a middle or late eighteenth century Historic Creek village. Several villages were known to have been located in the general area. Although it may have been part of Apalachicola, the site was not a large or important village.

A small, late Archaic component consisting of the typical, large, stemmed projectile points, fiber tempered pottery, and steatite sherds was also present. The content of the component is quite similar to the cultural material of the same period from 1 Ru 28 and probably represents the same phase.

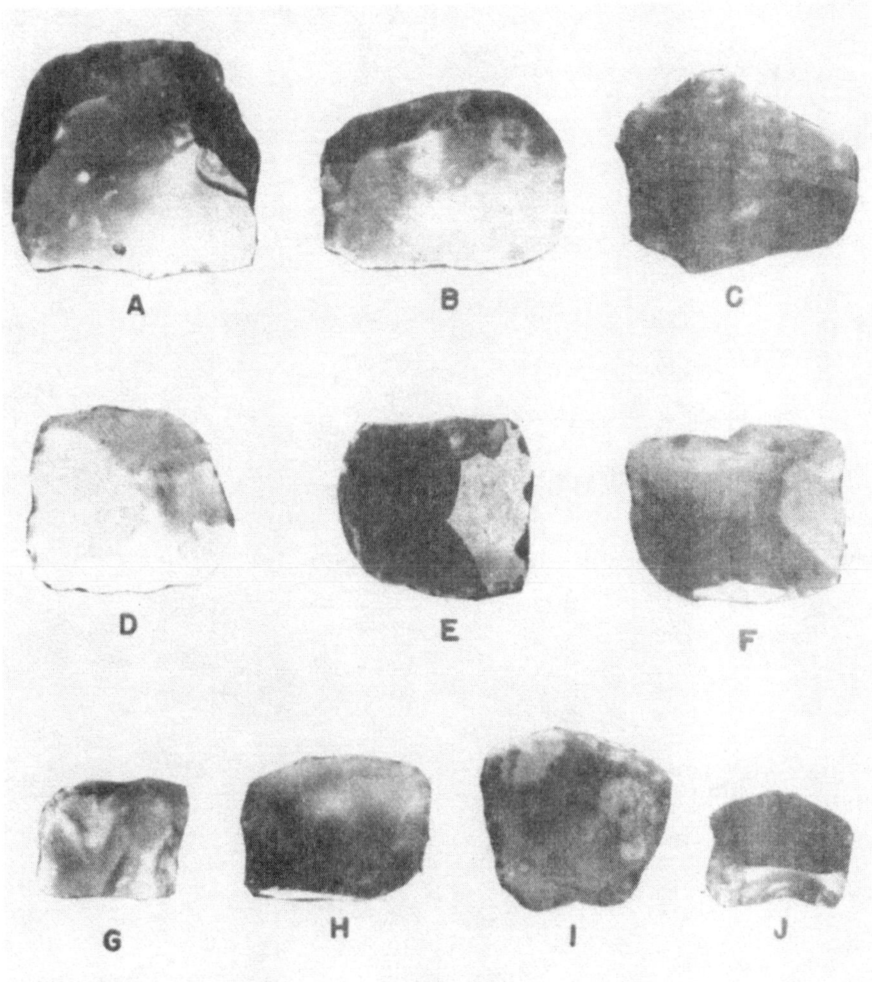

Figure 78 1 Ru 69 Gun Flints

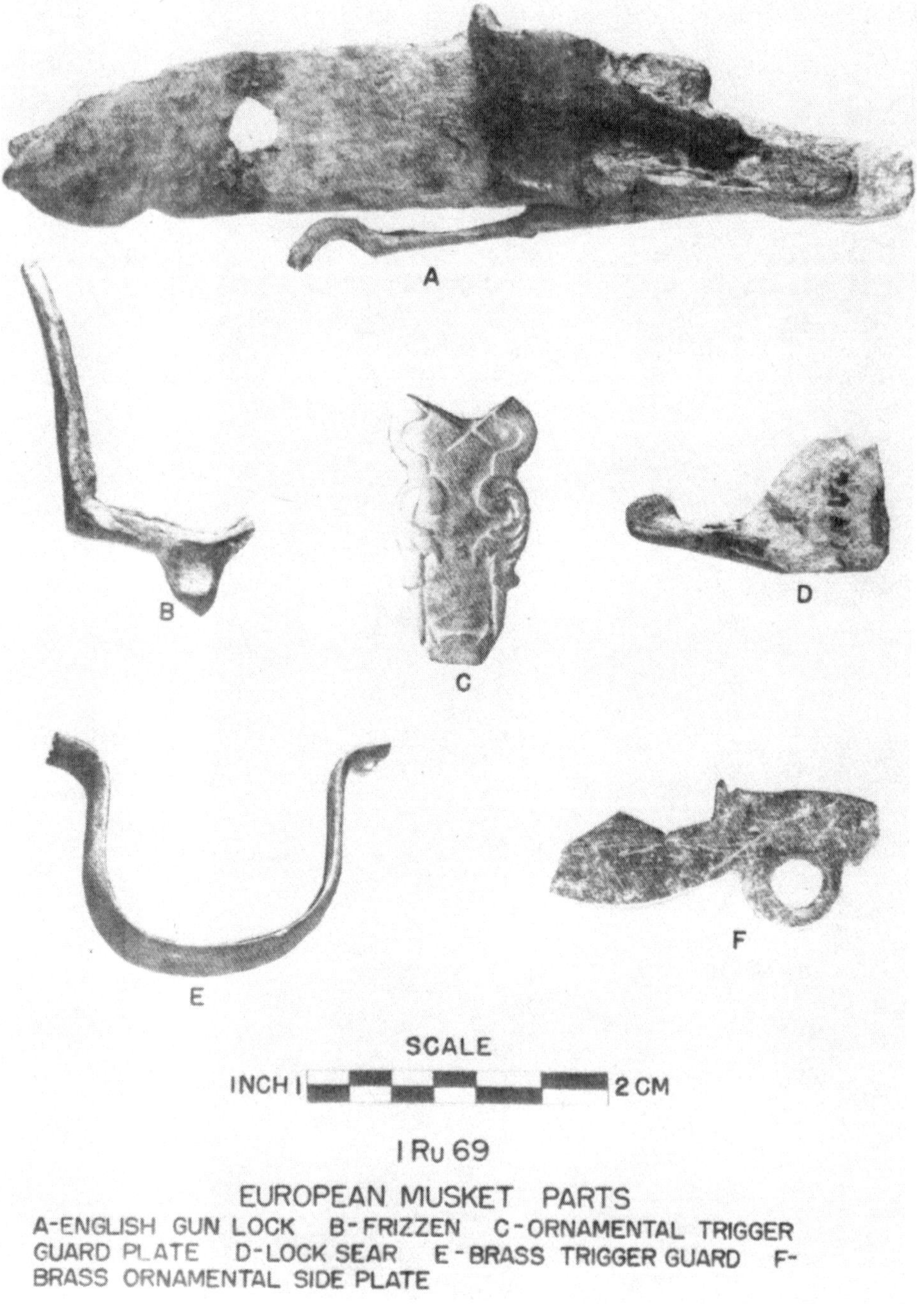

Figure 79 1 Ru 69 European Musket Parts

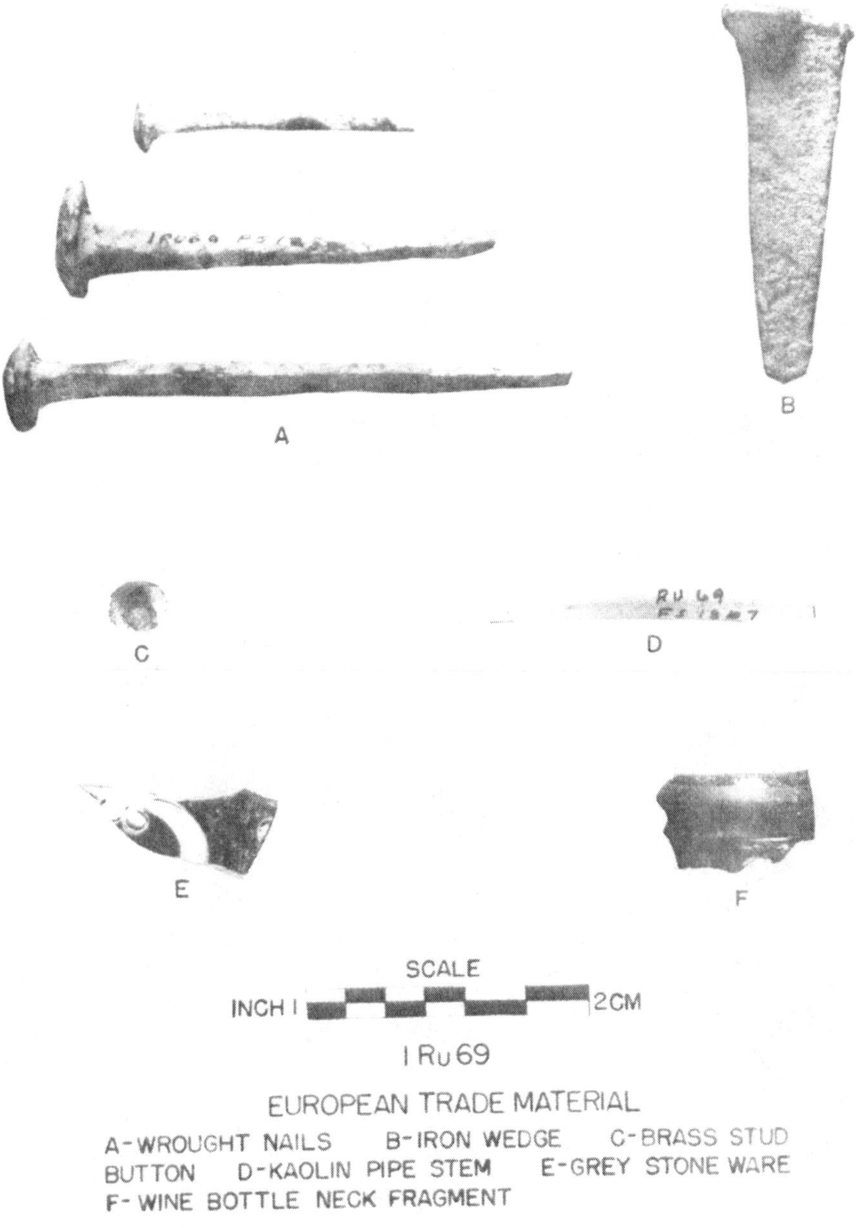

Figure 80 1 Ru 69 European Trade Material

CHAPTER 5

Summary and Conclusions

The sites at which the excavations reported in this study took place represent five major periods of culture history in the middle Chattahoochee Valley. Other culture periods of the area are described in reports of the intensive work that has been completed on other phases of the valley's prehistory by the Smithsonian Institution and the University of Georgia.

The long pre-ceramic occupation of the area is least known. Although a few widely scattered artifacts possibly representing this occupation have been found, little can be added to Hurt's statements concerning the Paleo-Indian period or the Lithic stage in the valley. Further interpretation of Hurt's work based on later findings is possible, however. One recalls that several projectile points found by him in private collections were of the Dalton type. The early status of the Dalton complex has now been well documented, as the Stanfield-Worley site seemed to have been transitional in several respects between Paleo-Indian and Archaic (DeJarnette, Kurjack, and Cambron 1962). Evidently the small groups whose occupations are marked by this artifact complex were both numerous and widespread. While a few artifacts that may have been part of the complex were encountered within the basin, Hurt's points were collected some distance from the Chattahoochee in the hills near Seale, Alabama. A preference for upland areas for the location of Paleo-Indian and Early Archaic settlements in the Oliver Basin to the north has been noted by McMichael and Kellar (1959).

Carefully made projectile points are the most distinctive artifacts found on Early Archaic sites, where they form a much greater proportion of the total assemblage than at later sites. This proportion indicates a dominance of hunting in economies of the period. While it cannot yet be determined whether the subsistence techniques of the fully Paleo-Indian peoples of the valley differed much from those of the Dalton, Old Quartz, Standing Bay, and other Early Archaic cultures, the latter economies seem to have been manifestations of the continuing earlier tradition.

The Late Archaic period is slightly better defined; traces of it can be found at many sites within the basin. Large, stemmed projectile points continue as a part of the assemblage, along with grinding stones and other suggestions of a vegetable gathering economy. The addition of steatite vessels and fiber tempered pottery to this complex seems to mark its climax; fiber tempered wares, steatite, and "Archaic-style" projectile points were found at most of the sites examined in the valley.

Although their economic activities differed, the early ceramics of Indians of the Chattahoochee Valley are apparently related to the fiber tempered wares of Indians of Stallings Island, for Stallings Punctated is a rare but consistent characteristic of the valley's fiber tempered pottery complex. Pottery tempered with vegetable fiber is a terminal addition to the Stallings Island Archaic complex, whose principal characteristic was its specialized economy based on the exploitation of shellfish. A more generalized economy including hunting, gathering, and fishing was present on the Chattahoochee. The general trend of the Shell Mound Archaic in the Southeast, involving increased wealth and increasing efficiency, seems to hold true for the Chattahoochee River Archaic as well.

The largest, most numerous sites of the Early Woodland period (so called for lack of a more exact designation) are those that are components of the Cartersville phase (Caldwell 1958). These sites are characterized by large amounts of check stamped pottery, slightly smaller amounts of simple stamped ware, and small numbers of cord wrapped, dowel impressed sherds. This definition is least adequate for early and late parts of the Cartersville phase, since the fabric marked sherds were the result of influence from areas further north, and the Chattahoochee Valley is considered marginal to the area of the middle eastern tradition where fabric marked types are a characteristic of the Early Woodland.

The difficulties in using check stamping as a diagnostic trait plagued Hurt's early study as well as studies by others. It is conceded that the check stamp tradition in the Southeast lasted from Deptford-Cartersville to Historic. Check stamping has been noted in Swift Creek, and indeed, the few Early Swift Creek Complicated Stamped sherds at 1 Br 15 were probably related to the later part of the Cartersville occupation rather than a later Swift Creek component.

Simple stamping, lasting into Swift Creek, also has had a long life in the valley. The earliest phase of the stamping tradition in the area is marked by a simple stamp complex called Mossy Oak. A wide range of simple stamped wares was present at 1 Br 15, but it was not possible to distinguish those that might have predated or antedated the Cartersville component.

Little concrete evidence of economic activity was found. Projectile points and other signs of hunting activities appear to decrease; yet definite evidence of a continued Archaic economy based largely on gathering of vegetable foods was not noted. Despite some hints of a formative economy at this early level (Morse 1960), the continuance of Late Archaic economic traditions suggested by McMichael and Kellar (1959) appears to describe the situation most adequately.

No components dating from the span of time between Cartersville and the first fully formative Mississippian manifestations were examined by the University of Alabama during the field seasons spent in the George Basin. Early and Late Swift Creek sites, with the Late Swift Creek sites containing increasing amounts of Weeden Island wares toward the south end of the basin, were reported by Hurt. Small amounts of Weeden Island sherds were found at various excavated sites. Although no Averett sherds were observed among the collected plainwares, some sherds must have been present in the valley.

Three phases of the Mississippian tradition were examined in the Chattahoochee Valley. According to existing terminology, these phases are known as Rood's, Bull Creek, and Ocmulgee Fields. The "Lamaroid" component at 1 Br 15 actually seems most closely related to the Early Mississippian component at the Rood's Landing site. More correctly, the 1 Br 15 component can be designated "late Rood's phase," for considerable Fort Walton and classic Lamar material is present.

A group of components resembling the classic Lamar follow the Rood's phase. The resemblance is closest at the north end of the valley; the Lamar component at Abercrombie, for example, is characterized by a high incidence of Lamar Complicated Stamped. Further south, Fort Walton wares increase while complicated stamped wares decrease. McMichael and Kellar (1959: 215) term this phase the "Bull Creek Focus" and define it as "Rood's Focus acculturated in the Lamar tradition, but with the addition of Fort Walton-like traits."

At least two subphases are present during the Historic period. They are easily seen by contrasting the Historic component at Abercrombie with the later, more typically Ocmulgee Fields sites. Operationally, large amounts of brushed, grit tempered wares are characteristic of late Historic ceramics, while shell tempered pottery including less brushed ware seems more characteristic of early Historic ceramics. How the early wares are related to cultures in other areas, especially cultures of the Upper Creek country and the Dallas and Mouse Creek cultures of Tennessee, has yet to be determined; both time and ethnic variation may account for some of the diversity noted during the early Historic period. McMichael and Kellar regard their Ocmulgee Fields I subphase as a complex of Bull Creek focus remains plus several additions.

McMichael and Kellar note a trend toward homogeneity in their Ocmulgee Fields II. These sites represent the remains of the well-known Lower Creek towns mentioned in ethnographic accounts. Goods of European manufacture are common on these sites, though scarce on earlier. Perhaps the most distinctive artifacts of the later sites are English claypipe fragments.

Part III

SPECIAL INVESTIGATION OF 1Ru 101, THE SPANISH FORT SITE

Edward B. Kurjack and Fred Lamar Pearson, Jr.

CHAPTER 6

Historical Background

The appearance of "Two Early Letters from Alabama" in *The Alabama Review* (1956) inaugurated a search to find the location of Fort Apalachicola, a Spanish fortress constructed on the Chattahoochee River in the late seventeenth century. These two letters from the then commandant, Lt. Favian de Angulo, to the Spanish governor of Florida describe both the fortress and its environs. Editor Mark Fretwell suggested in his article that the fort stood in Russell County, Alabama, "between the mouths of Ihagee Creek and Snake Creek" and specified an area "enclosed by a twisting looping bend of the Chattahoochee" about two miles southeast of Holy Trinity. Brother Finbar of the Trinitarian Mission there soon located an earthwork that appeared to fit the description of the fort. This earthwork is located on the north side of a bend in the Chattahoochee at an elevation of about 210 feet and adjacent to a 60-foot bluff overlooking the river. Harold Huscher of the Smithsonian River Basins Survey examined the site in May, 1958, noting its importance as an historical monument and as an archaeological horizon marker. In the summer of 1960 as part of a cooperative archaeological survey of the George Basin by the United States Department of the Interior-National Park Service and The University of Alabama, Lewis Larson conducted initial excavations at the site, now designated 1 Ru 101, in an attempt to determine if it were indeed Fort Apalachicola.

Spain entered the Western Hemisphere in 1492 and soon established supply bases and permanent colonies on the islands of Cuba, Española and Puerto Rico. Six major expeditions to Florida were launched between 1513 and 1562, but with such little success that in 1562 Philip II ordered that no further expeditions be sent to Florida. Spanish interest quickly revived when the king learned that the French Huguenots had colonized at the mouth of the St. Johns, posing a serious threat to the Bahama Channel, the return route for the Spanish silver fleet. French vessels based in the river could easily prey on the treasure ships and return to a safe harbor. The Spanish government sent Pedro Menendez de Avila in 1565 to colonize Florida and counteract the French threat. Menendez established defensive outposts on the coast and in the hinterland. Missionary activities accompanied the military, and by 1572 outposts dotted the Atlantic seaboard as far north as Chesapeake Bay. During the sixteenth and seventeenth centuries the Caribbean presented a prime hunting ground for Dutch, English, French, and even Spanish pirates. After the establishment of Jamestown in 1607 England replaced France as Spain's chief rival for the control of the Atlantic coast. Spain's military interest was concentrated on the seacoast, but the Spanish Franciscans, spurred by the Church in Cuba, pushed westward to begin missionary work among the Apalachee. While the Franciscans were concentrating on the erection of Apalachee missions in west Florida, Anglo-Spanish rivalry to the east was intensified by the establishment of Charleston in 1670.

English planters, principally from Barbados, left the West Indian islands in search of new farmland and increased acreage. The settlers at Charleston began to apply pressure to Spanish Georgia, with the result that the Spanish began a gradual retreat down the Atlantic seacoast. English influence not only expanded southward along the Georgia seaboard but also spread west toward Alabama through the activities of Carolina merchants. Chief among these was Dr. Henry Woodward of Charleston, whose influence brought the Westos into the South Carolina Indian Alliance of 1674-1680. Woodward's alliance not only served as an important defense against the Spaniards but also opened up the way for English penetration of the Georgia interior. Although by 1682 the greed of Carolina slave traders had broken the Indian alliance and diminished the once powerful Westos to a tribe of hardly fifty, and although Henry Woodward escaped the Spanish soldiers only to die of wounds after a hostile Indian ambush, the English appearance on the Chattahoochee presented a serious threat to Spain's control of western Georgia and Florida. She recognized it as such. Spanish colonial authorities determined that the Chattahoochee must be fortified and that the English advance must be checked, for to lose control of the Chattahoochee would leave the Apalachee missions exposed to attack. Governor Diego de Quiroga y Losada of St. Augustine visited Apalachee province in 1687. He found the Apalachicolas still smarting from earlier punitive Spanish expeditions—four of their towns had been burned because they had concealed and aided English traders. Quiroga assured them of Spain's desire for their friendship and trade. When Quiroga's aid in their resettlement did not prevent the Apalachicolas from trading with the English, the governor announced plans to build a *presidio* on the Chattahoochee and to station infantrymen in the province to keep the traders out. Officials pointed out the difficulty in maintaining a garrison so far from established posts (160 leagues from St. Augustine and 80 leagues from Apalachee), but he decided nonetheless to build in 1689 at the town of Apalachicola. Quiroga chose Capt. Primo de Rivera to construct the *casa fuerte* because of his experience with fortifications while serving in Flanders. Governor Quiroga sent 100 Indians, many of whom were carpenters, with Captain Rivera so that the fort could be completed within two months. The finished fortress, complete with bastions and moat, was structurally similar to the Castillo de San Marcos at St. Augustine. Lt. Favian de Angulo traveled up the Chattahoochee in the spring of 1690 to assume command of the garrison. Two of his letters to Quiroga while stationed at the fort survived to appear in *The Alabama Review.* He reported on April 14 that the garrison consisted of a corporal, two *reformados* (officers deprived of former commands), seventeen regular soldiers, and twenty Apalachee Indians. His letters described the fort itself and also a conference with the Indians. Angulo did what he could to maintain friendly relations with the Indians. He gave them seed corn and promised to supply them with garrison stores in time of need. At Caveta, Angulo told the Apalachicolas that they had erred in trading with the English. The Indians replied that the English had supplied them with guns, ammunition, and other needed articles. When Angulo told them the garrison was to protect them from the English and from hostile Indians and that they must report to him if the English resumed trading in their villages, he also demanded that they make their allegiance clear. Many, no doubt, remembered the burning of their towns that had resulted from their protection of Woodward, and they affirmed their loyalty to Spain.

With the building of the fortress of the Chattahoochee in 1689 and the strengthening of its garrison the following year, Spain seemed to have gained the upper hand over her European rival. The success proved short-lived, for the fortress lost its importance after less than a year. The Apalachicolas were convinced that the Spanish garrison would prevent the return of the English fur traders and therefore cause the loss of English trade items, which were of superior quality and in greater supply than those available through the Spanish. Probably because of this desire for English trade, most of the Apalachicolas left to relocate on the Ocmulgee River in central Georgia. Fort Apalachicola no longer served a useful purpose.

In 1691 Juan de la Frente reported in Havana that French pirates planned to attack St. Augustine. Governor Quiroga, receiving no outside help in strengthening the *presidio,* saw that men and supplies would have to be brought from other sections of Florida. The garrisons at Guale, Apalachee, and Timucua were currently subject to pirate and hostile Indian raids. Therefore, knowing that most of the Apalachicolas had left the Chattahoochee, Governor Quiroga dispatched an order to the fort commandant, Lt. Patricio de Florencia, with specific instructions to demolish the fortifications, fill in the moat, and remove arms, food, and other supplies in order to make the site useless to the Carolina merchants. At the same time Florencia was to invite the departed Apalachicolas to reoccupy their deserted villages provided they would no longer trade with the English. Lieutenant Florencia and the garrison carried out the governor's orders, then left to reinforce the *presidio* at St. Augustine.

Fort Apalachicola represented a high watermark of Spanish expansion in the Georgia country. The Anglo-Spanish rivalry involved missionaries, soldiers, and traders. Spain's Indian policy was conducted primarily by the padres and soldiers, while that of England was carried out principally by men of commercial interests. Spanish expeditions to stop the English intruders failed mainly because the Indians desired English trade goods. The construction of Fort Apalachicola, by triggering the movement of the Apalachicola Indians in the direction of Charleston, probably helped the English instead of hindering them.

CHAPTER 7

Excavations

The Fort Apalachicola site is now a flat, rectangular mound, surrounded by a shallow ditch. Trianguloid projections, immediately reminiscent of the bastions shown in Quiroga's map of the fort (see Figure 88), are found at each corner. The shape and size of the site are indicated on the contour map, Figure 81.

Two trenches were excavated at the site. The largest of these, extending from the interior of the fort to its approximate center and crossing the outside earth embankment and moat, was located near the middle of the eastern side of the fort. This first trench was enlarged to examine various structural features. The second trench consisted of five squares excavated just outside the palisade wall in the center area of the northerly side of the site. All excavation was done by 5-foot horizontal units and 6-inch levels. Natural stratigraphy was utilized in the structural features, which are described below.

STRUCTURAL FEATURES

The Trace. The fort trace, complete to the corner bastions, was visible on the ground surface and outlined by the slight moat depression. Angulo's letter to Quiroga listed the dimensions of the fort as 22 *varas* north-south, 19 *varas* east-west. The *vara,* equivalent of the English yard, varied in length from approximately 30 to 35 inches. Fretwell (1956) gives these lengths as 61.16 feet north-south and 52.82 feet east-west. The dimensions probably indicate the size of the rectangle formed by the palisade walls. The remains of the palisade, described below, were located inside the moat. A projected outline of the fort based on Angulo's dimensions and the configurations of the early plans will, when aligned with the uncovered 15 feet of the palisade, fit within the moat as it appears on the present contour map. Small discrepancies can be attributed to the indeterminate length of the *vara.*

The Moat. Angulo recorded the size of the moat surrounding the fort as 4 *varas*

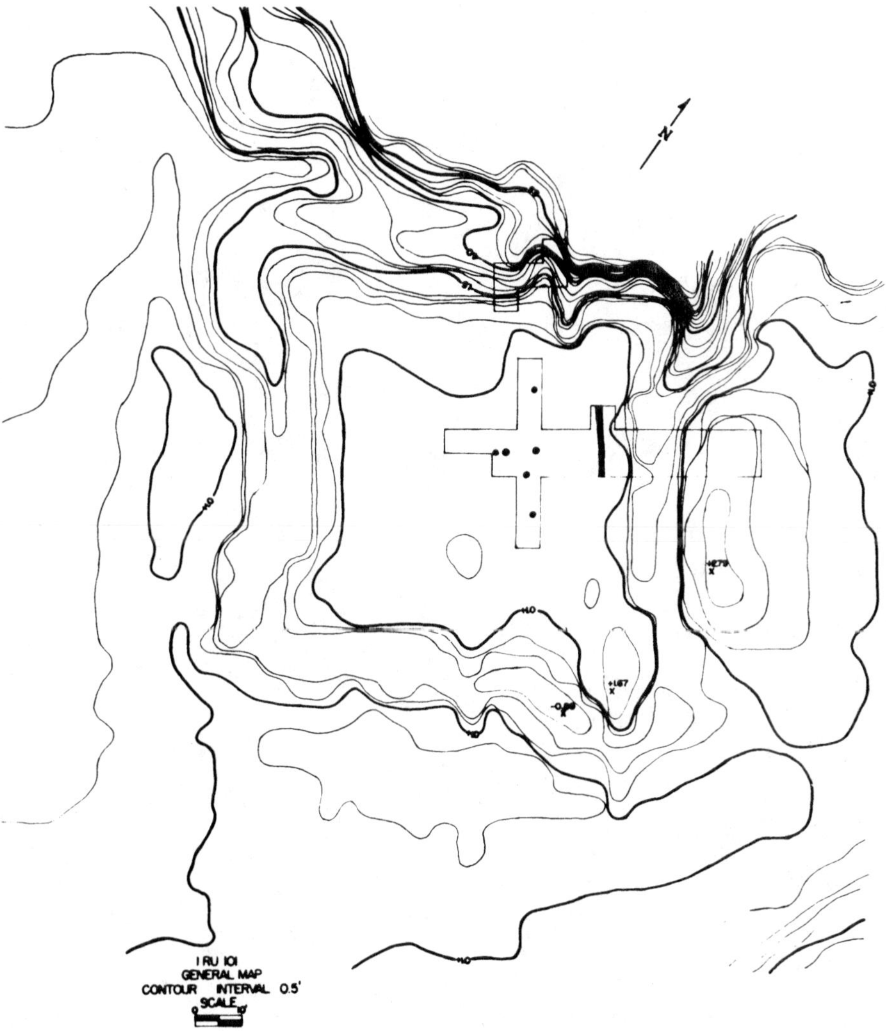

Figure 81 1 Ru 101 General Map

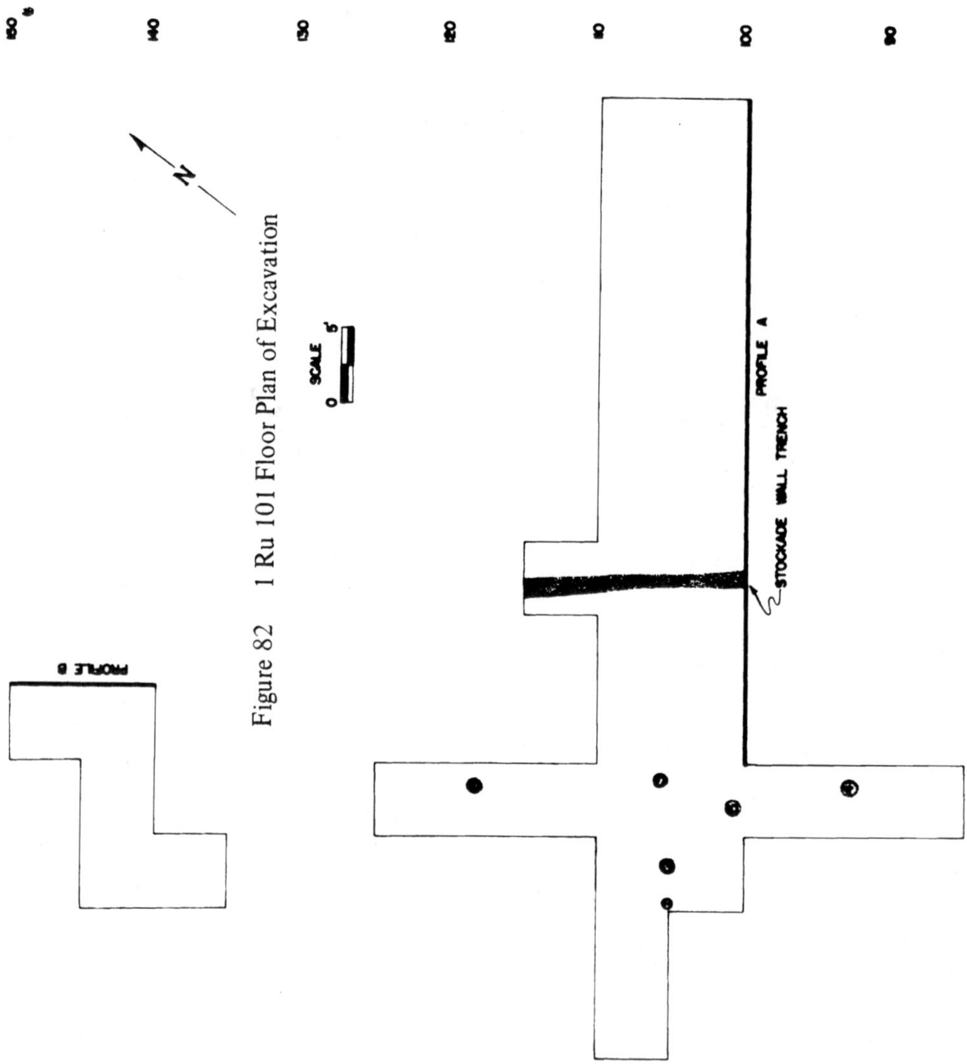

Figure 82 1 Ru 101 Floor Plan of Excavation

Figure 83 Excavations at 1 Ru 101

Figure 84 1 Ru 101, Trenches 100 L 70, 100 L 75, 110 L 80, and 110 L 85

Figure 85 Site 1 Ru 101, Squares 110 L 50, 110 L 55, 100 L 50, and 100 L 55

Figure 86 Site 1 Ru 101, Looking Southwest in Trenches 100 L 85, 100 L 80, 100 L 75, and 100 L 70

across, translated as 11.12 feet. The profile drawing (Figure 87, Profile A) shows a cross section of the moat, which was filled with clay, charcoal, and fired daub. The fired daub consisted of large, burned, clay fragments, often a foot or more in diameter and usually containing some vegetable material. The top layer of the moat was composed of soil that might have been washed into the depression, but the soil layer is consonant with Bolton's suggestion that the moat was filled in when Fort Apalachicola was abandoned. Areas of what seemed to be finished surface appeared on some of the daub fragments, and the impressions of posts appeared on others. Large pieces of charred wood were particularly numerous near the base of the daub layer.

Angulo's description of the palisade indicates that it was of wattle and daub construction, reinforced by an exterior half wall of clay. After the fort was destroyed to prevent use by the English, the remains of the parapet must have fallen.

Earth Embankment. Outside the moat, the remains of an earth embankment were evident on the surface and in the profile of the excavation. The surface in the area of this embankment was occasionally higher than the surface in the interior area of the fort, indicating that the embankment may have lost much of its height when the moat was filled.

Remains of the Stockade Wall on the North Side of the Fort. Evidence of a footing trench for a stockade wall was clearly visible in the area excavated inside the moat along the edge of the fort (Figure 87, Profile A). The footing trench exhibited vertical sides and a relatively flat bottom. It measured a foot in thickness. The limited excavations along the exterior edge of the fort on the north side revealed a fired daub layer that seemed to be the remains of the fallen stockade wall. Its situation is shown on the profile drawing (Figure 87, Profile B).

ARTIFACTS OF EUROPEAN ORIGIN

Two groups of artifacts made by Europeans were collected in the fort.

The first group, consisting of majolica sherds and olive jar fragments, was the only set of objects that could definitely be classified as Spanish. One of the majolica sherds was of the type *San Luis Blue on White* (Goggin, Ms., 1957). This type dates from a period overlapping the period of occupation of Fort Apalachicola. Because of damage (apparently the result of fire) to their finish; the rest of the majolica sherds were not classifiable.

According to John Goggin, who examined the specimen, the olive jar sherds were made from a gritty paste fitting the description of a ware used during the "Middle Period" (about 1580 to 1780). The few large body sherds appeared to be from "shape B" vessels (Goggin, 1961).

The second group of European artifacts appeared to date from the latter part of the eighteenth century. All of the wine bottle fragments, including a bottle seal, were classified into this category. The bottle seal, 1.2 inches in diameter, was impressed on light green glass and bore the name "Lafitte Medoc." The seal was identified by Paul Hudson (July, 1960) as definitely dating from the late eighteenth to early nineteenth century.

ARTIFACTS OF INDIAN MANUFACTURE

Pottery. Many sherds were found in all areas of the fort. (Even unusually large vessel fragments were not uncommon.) One almost restorable vessel, illustrated in Figure 91, was also collected. The frequencies of the various pottery types are shown in Table 30. Sherds from the site were classified according to the following ceramic types and sorting categories.

Ocmulgee Fields Incised (Jennings and Fairbanks, 1939). Both forms of this ware—casuela bowls incised on the exterior of the vessel above the shoulder, and plates incised on the interior side of the flaring rim—were present. Numerous incised, rectilinear and curvilinear motifs, which sometimes enclosed rows of punctations, were present. The thin incision lines associated with this type of pottery were present, and many sherds found at the site showed smoothed-over incising. A few sherds classified as plate fragments suggested round-bodied bowls with incised, flaring rims. Extruded or thickened lips were present on the rims of casuela bowls; infrequently, the extrusions were notched or underlined by a row of punctations. Notched or punctated decorations along the shoulder of casuela bowls were present but rare; a notched fold under the rim on the exterior of the plates was more common. Almost all of the Ocmulgee Fields Incised sherds were tempered with fine shell and grit, and could be graded from sherds with large amounts of finely crushed shell to sherds having little or no shell in the paste (however, shell was visible in most of the pottery fragments). The ware characteristics of most of the Ocmulgee Fields Incised sherds differed from characteristics of sherds classified in the brushed types. Few of the basal portions of these incised vessels were brushed or roughened.

Incised Round-Bodied Bowls. Several rim sherds and the nearly complete vessel illustrated in Figure 91 were parts of plain or brushed, round-bodied vessels decorated with a crudely incised band just below the rim. The lips of these vessels were often adorned by a small, pinched or notched, rim fold. All of the sherds in this group were tempered with large amounts of coarse, crushed shell.

Walnut Roughened (Anon. 1940a). The brushed and cobmarked sherds tempered with coarse shell were round-bodied vessels that usually displayed unroughened areas adjacent to the rim. A few fine shell tempered, brushed sherds were collected; most of them were lightly brushed.

Chattahoochee Brushed (Bullen 1950). A few brushed, coarse grit tempered body sherds were classified Chattahoochee Brushed.

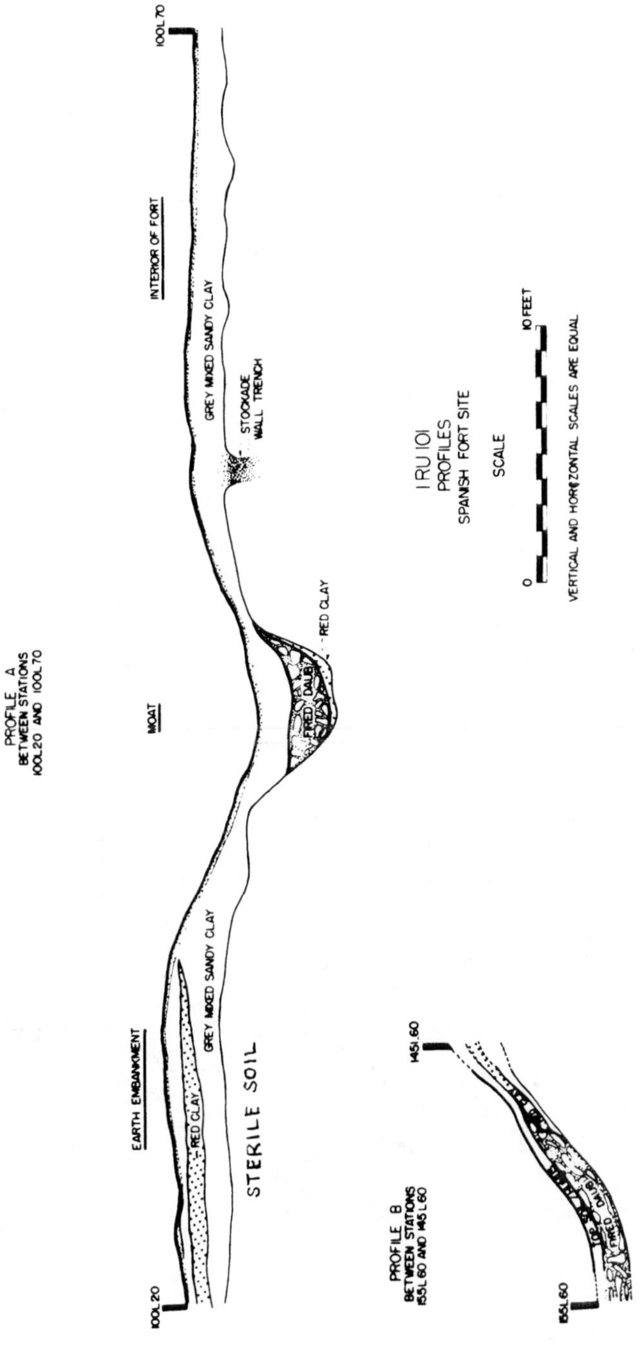

Figure 87 1 Ru 101 Profiles, Spanish Fort Site

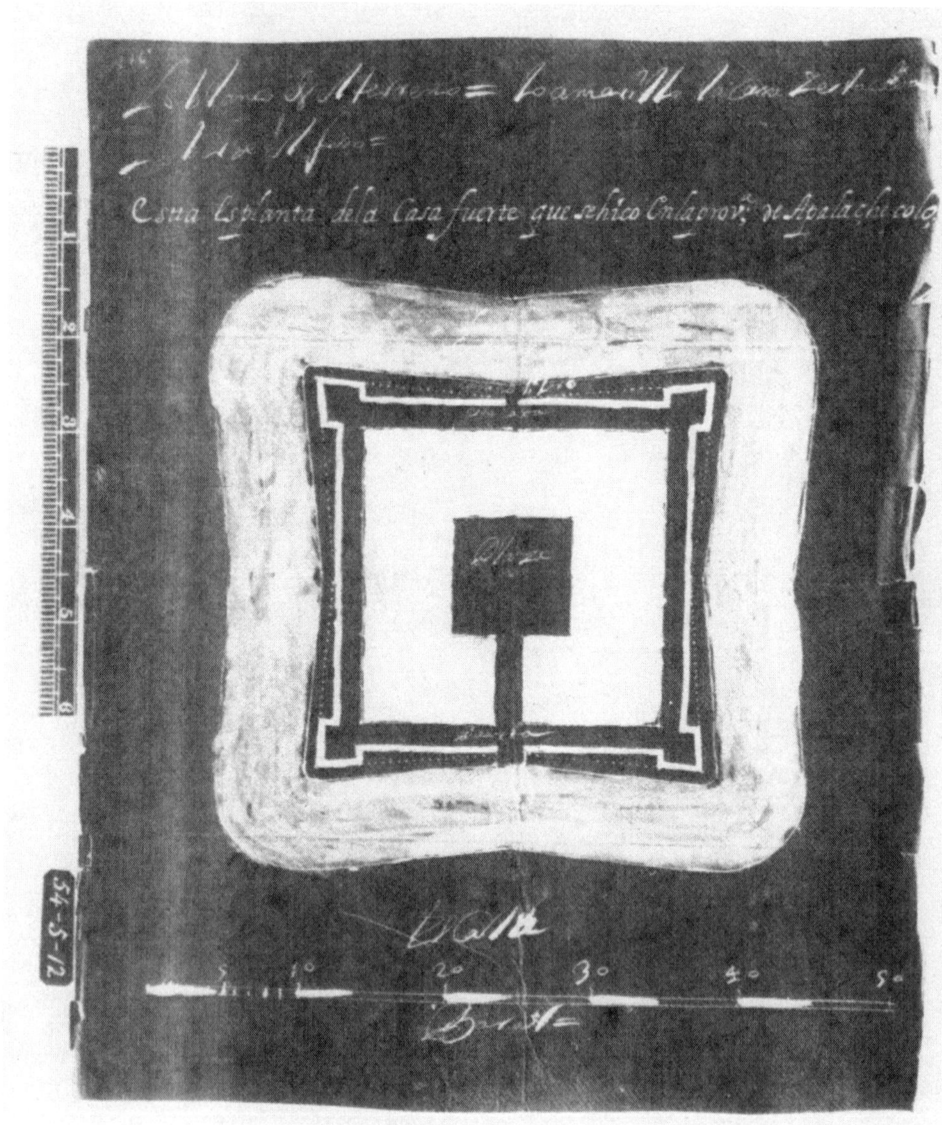

Figure 88 Map of Fort Apalachicola

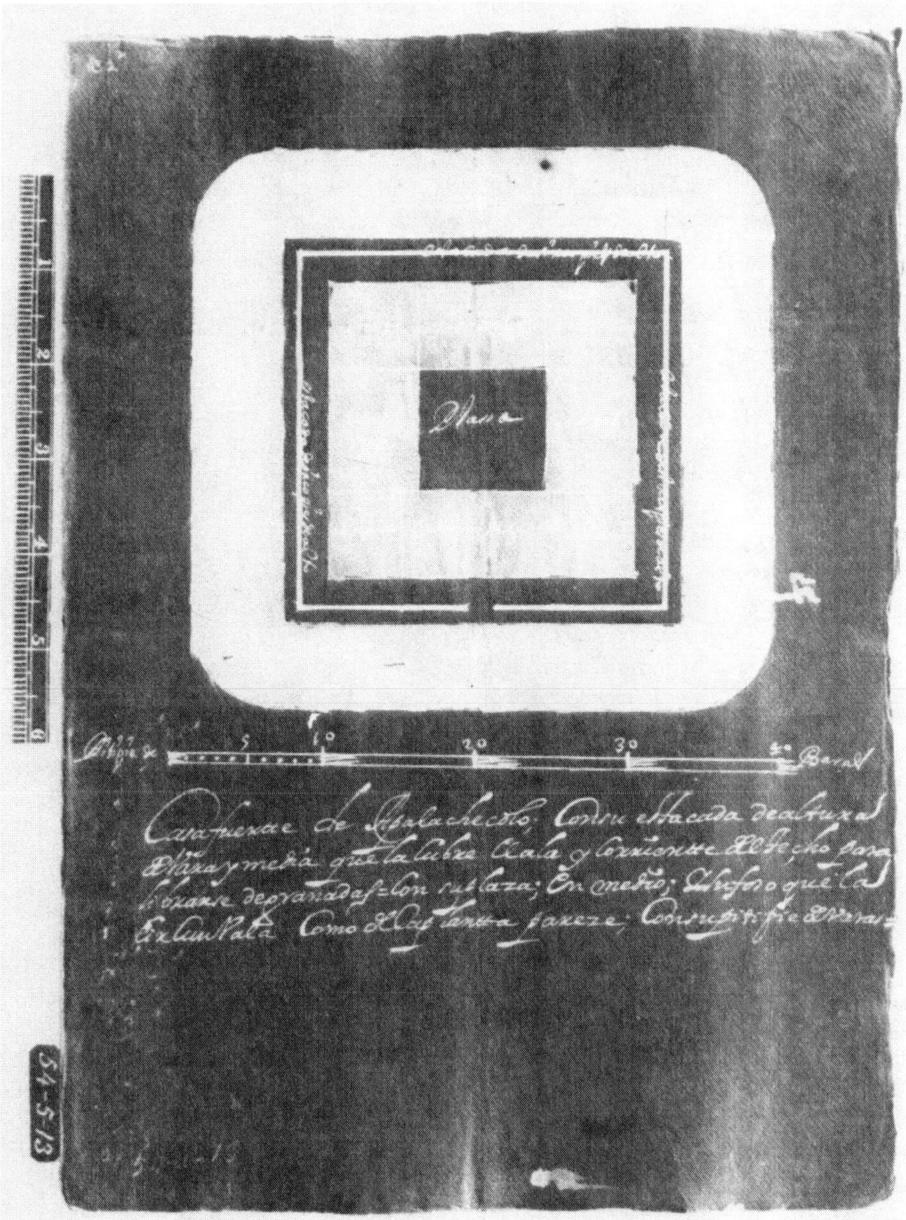

Figure 89   Later Map of Fort Apalachicola

Kasita Red Filmed (Haag 1940b). A few small sherds of this type were present. The red filming was found in zones that were often outlined by incised lines. All of the sherds suggested a shallow plate form with decorations on the interior. One example of annular ring at the base of a plate was found. The rims of the vessels were slightly folded or extruded on the exterior. The sherds were tempered with fine shell and grit.

McKee Island Cord Marked (Heimlich 1952). A single, shell tempered body sherd, roughened by the random application of a paddle wrapped with fine cord, was collected. McKee Island Cord Marked is an important McKee Island complex type; its contemporaneity with historic European trade materials and other McKee Island pottery types is demonstrated by burial associations.

Lamar Complicated Stamped (Jennings and Fairbanks 1939). A few coarse grit tempered, complicated stamped sherds were collected.

Coarse Shell Tempered Plainware. Most of the sherds from this category were parts of plain, round-bodied, utilitarian vessels or were the plain parts of brushed vessels. Rims of this ware were usually slightly flaring with small, pinched or notched, exterior folds. Several examples of strap handles were noted. The very few vessels of this type that were decorated by crude, incised designs just below the rim, were classified in the Incised Round-Bodied Bowl category.

Coarse Grit Tempered Plainware. A few coarse grit tempered sherds were found, but yielded no information as to vessel form.

Fine Shell and Grit Tempered Plainware. Sherds classified in this group included half of the entire ceramic collection from the site. Most of the sherds were undecorated fragments of incised, casuela bowls and plates; however, a few casuela bowl rim sherds were present.

Fiber Plain (Sears and Griffin 1950a). A single, fiber tempered body sherd was collected.

SUMMARY

Three vessel forms were used by the Indian potters supplying the fort: casuela bowls, plates, and round-bodied (globular) bowls. Most of the casuelas were incised above the shoulder and undecorated below. Shallow Kasita Red Filmed plates with annular basal rings were made, as well as the deeper Ocmulgee Fields Incised plates. A few undecorated examples of both plates and casuelas were also collected. The round-bodied (globular) bowls possessing slightly flaring rims decorated with pinched rim folds or appliquéd strips seem to have been made of an inferior, coarser ware. Most of these pots were plain, brushed, or cob marked. A few examples of similar vessels with incised designs in a band

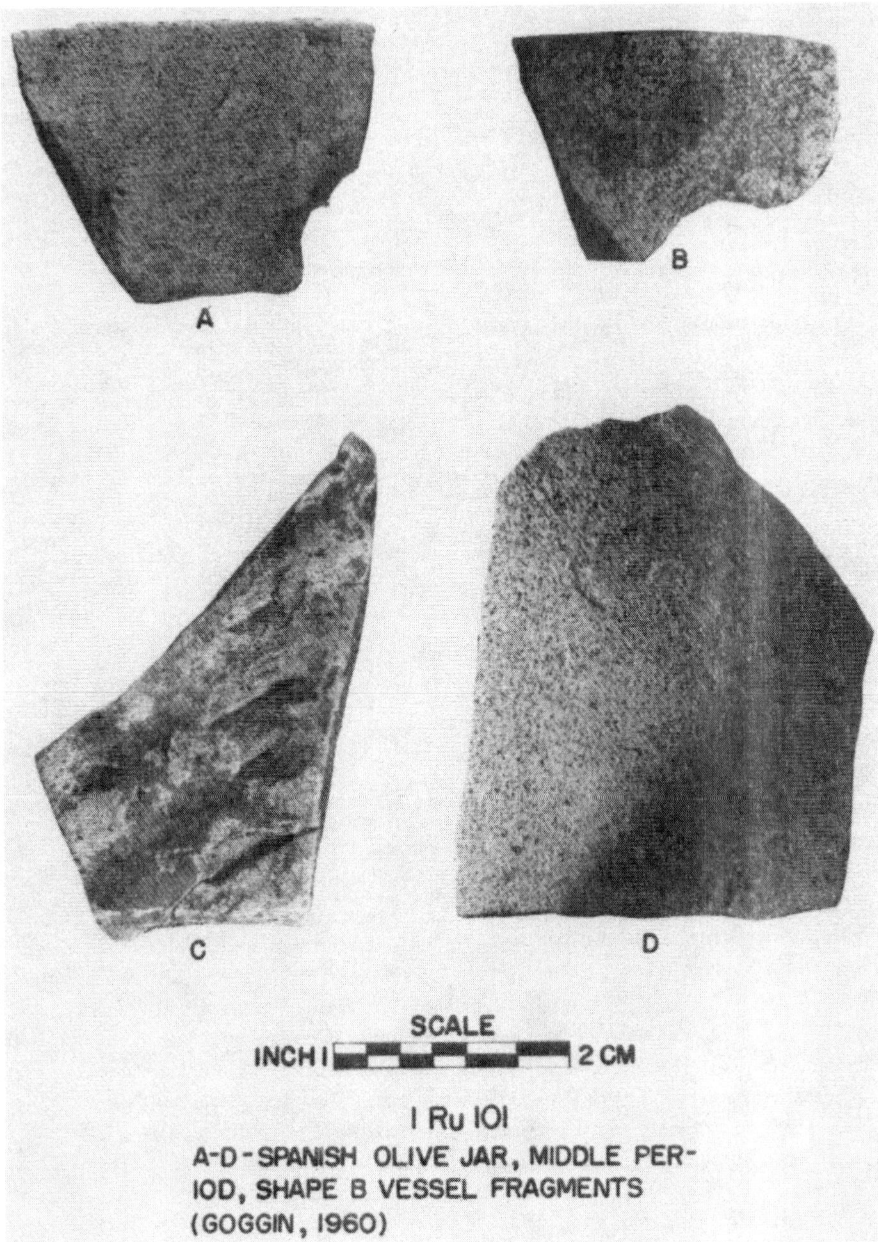

Figure 90 1 Ru 101 Spanish Olive Jar Sherds

TABLE 29

Artifacts of European Origin, 1 Ru 101	
Artifacts of Spanish Origin	
Olive jar sherds	7
San Luis Blue on White Majolica sherd	1
Re-fired Majolica sherds	3
Late 18th – Early 19th Century Objects (Probably Anglo-American)	
Green glass wine bottle fragments	6
Bottle seal	1
Ironstone sherds	2
Blue Feather Edge sherd	1
White Chinaware sherds	2
Brick fragments	1
Miscellaneous European Artifacts	
Ichtucknee Blue beads	2
Georgia Milk Pentagonal bead	1
Brass spangles	1
Wrought iron nails	5
Iron hinge fragments	2
Ox shoe (?)	1
Miscellaneous metal fragments	6

below the rim were noted, however.

Sherds from the round-bodied, coarse shell tempered, "utilitarian" vessels were relatively few in number, composing about a third of the entire ceramic collection. The ratio of incised sherds identified as casuela fragments to those identified as plates is approximately four to one. If most of the fine shell and grit tempered sherds are considered body fragments from the incised vessels, this ratio indicates that over half of the total pottery vessels made by Indians were casuela bowls. The incised plates constituted a tenth of the vessels. The Kasita Red Filmed plates were minor in proportion.

Ceramic Pipes.

Fourteen pottery pipe fragments were found at the site. Two of these are illustrated in Figure 93.

Figure 91 1 Ru 101 McKee Island Incised Vessel

TABLE 30

1 Ru 101 – Sherd Count

Type	Percent Type	Num-ber	Percent Total
Ocmulgee Fields Incised			
Casuela form	70.06	468	9.62
Plate form	13.27	102	2.10
Form not identified	14.67	98	2.01
Total		668	13.73
Walnut Roughened			
Coarse shell tempered, brushed	74.37	235	4.83
Coarse shell tempered, cob marked	24.37	77	1.58
Fine shell and grit tempered, brushed	1.27	4	.08
Total		316	6.49
Chattahoochee Brushed		15	.31
Lamar Complicated Stamped		1	.02
Kasita Red Filmed		18	.37
Fiber Plain		1	.02
McKee Island Cord Marked		1	.02
Incised Round Bodied Bowl		7	.14
Fine Shell and Grit Tempered Plainware		2,443	50.21
Coarse Shell Tempered Plainware		1,298	26.68
Coarse Grit Tempered Plainware		90	1.85
Fine Sand Tempered Plainware		8	.16
Site Total		4,866	100.00

Projectile Points.

Seven projectile points, five of them small and triangular, were collected at the site. The sixth was a thick, trianguloid projectile point; the seventh was a large, stemmed, serrated point. In addition, the midsection of a large projectile point was found.

Miscellaneous.

Two small scrapers, a steatite sherd, a polishing stone, and a sandstone atlatl weight were collected in the excavation. Most of these objects were not associated with the activity at the fort and represented earlier components in the area. They indicate clearly an aboriginal occupation not only prehistoric but preceramic.

FAUNAL REMAINS AND ETHNOBOTANICAL SPECIMENS

A considerable amount of animal bone was present at the site. Most of this material represents the remains of large mammals (most of them probably deer). Drumfish teeth were present as well as several mussel shells.

Two varieties of seeds were identified in the field: charred peach pits (12 whole and fragmentary pits were collected) and several pieces of charred corncob.

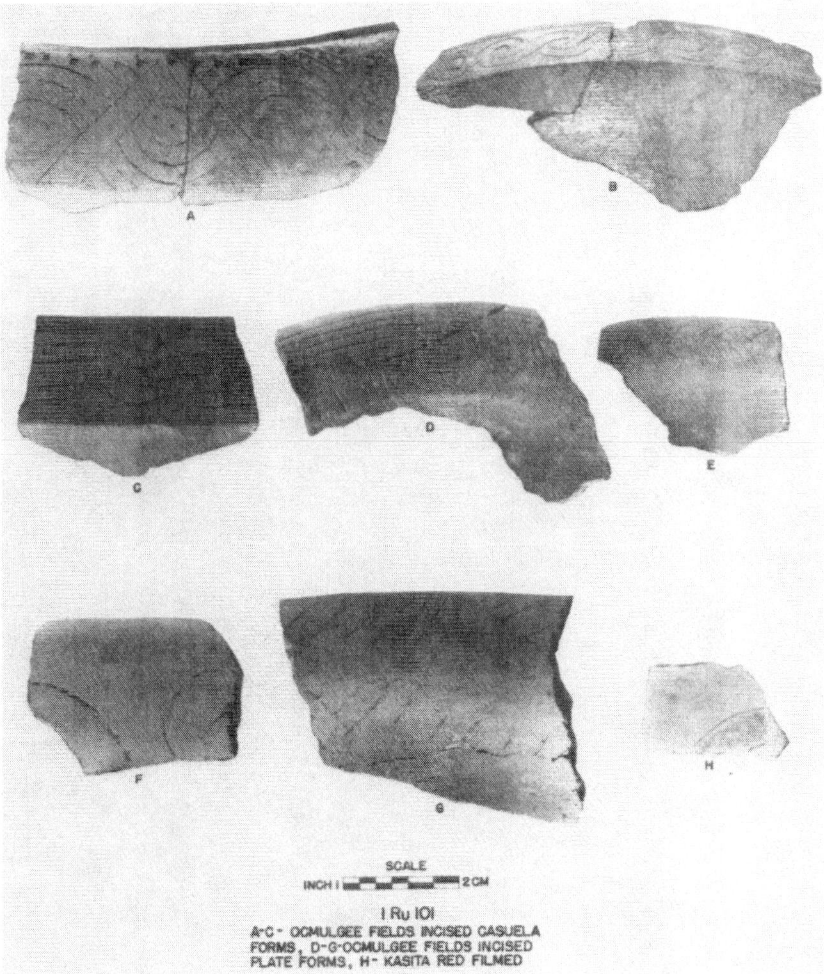

Figure 92 1 Ru 101 Miscellaneous Decorated Pottery Types

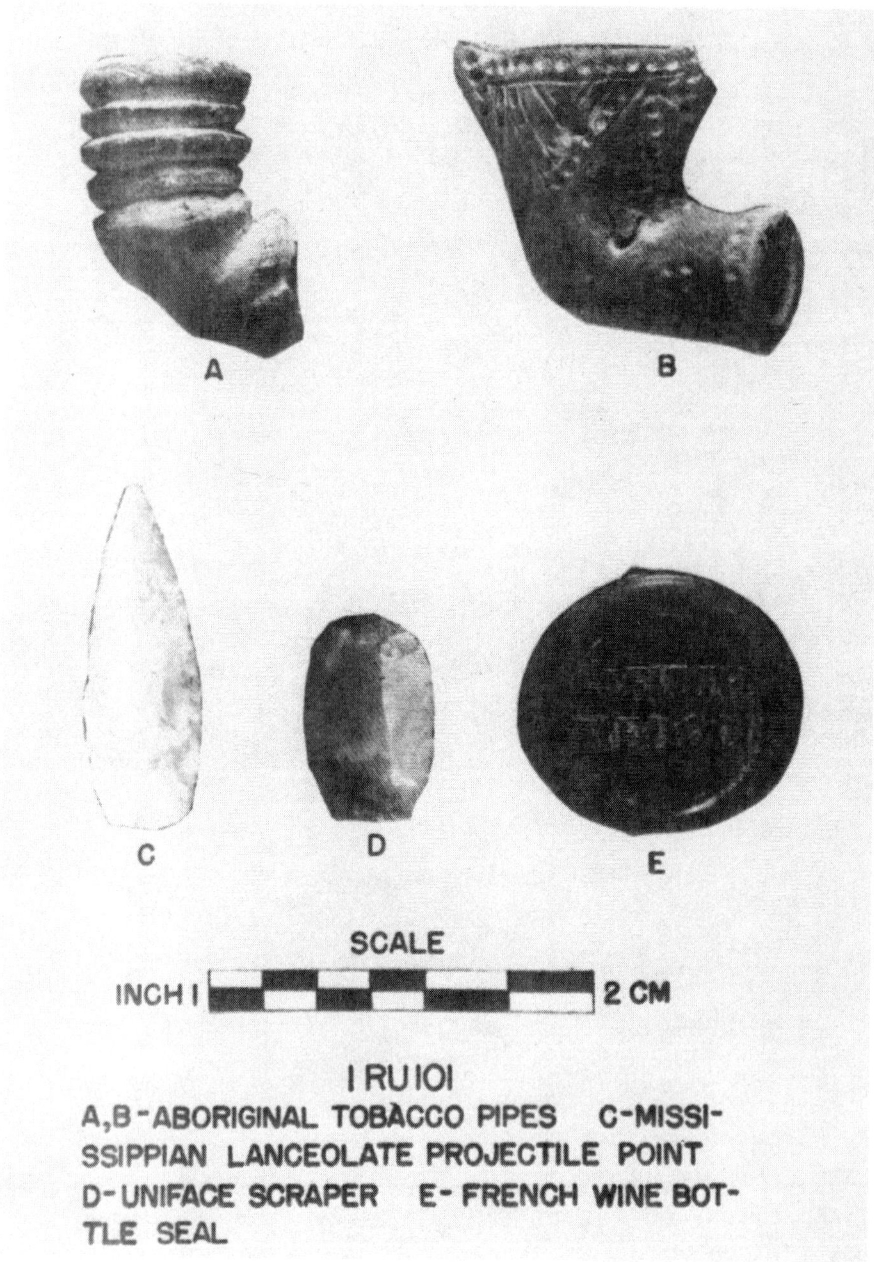

Figure 93 1 Ru 101 Miscellaneous Artifacts

CHAPTER 8

Summary and Conclusions

The principal purpose of the excavation at site 1 Ru 101 was to determine, if possible, whether the earth embankment located by Brother Finbar was the Spanish Fort Apalachicola that Captain Primo de Rivera built on the Chattahoochee in 1689.

Evidence–documentary, geographical, structural, and artifactual–strongly indicates that it was. First of all, the site is definitely in the area of the Fort Apalachicola described in Spanish documents of the late seventeenth century. Second, the sizes and shapes of the excavated structural features are in substantial agreement with Angulo's descriptions of the fort or *casa fuerte* that he commanded. Furthermore, the condition of the fort remains and the content of the filled moat are such as might be expected under the stated conditions of demolition and abandonment. Also, although the European artifacts include only a small amount of Spanish material, all of it is consonant with the dates of Fort Apalachicola. The scarcity of Spanish material is in keeping with the short occupation and the deliberate, careful removal of all usable articles. More remains would be expected under other circumstances, such as conquest or abandonment under attack. While there are a few European artifacts dating from a later period, which could be associated with an eighteenth and early nineteenth century occupation of the extensive village site around the fort, the scarcity and nature of the material preclude the site's having been an eighteenth or nineteenth century fort.

Equally important is the fact that while the Indian remains found at the site date from Archaic times to Mississippian, the great majority of aboriginal material can be placed in the late seventeenth century context of Fort Apalachicola, indicating a heavy occupation at this time.

Fretwell wrote:

In 1716, twenty-five years after the fort was destroyed, Diego Pena conducted an expedition from St. Augustine to the Indian towns along the Chattahoochee. In his journal he carefully recorded distances and described the country through which he travelled. In crossing to the west side of the river, he apparently arrived at a landing point near the northern boundary of the present Barbour County, Alabama and journeyed a short distance up the western bank to the village of Sabacola. The following night was spent "at a small farm . . . of the Casique of Apalachicola . . . two leagues, more or less," distant from the main village of Apalachicolo.

Pena added, "Today there is found a small farm, where once there was a fort."

William Bartram also visited these river towns. . . . Of his sojourn there in 1777, he wrote:

> One day the chief trader . . . obliged me with his company on a walk of about a mile and a half down the river, to view the ruins and site of the ancient Apalachucla: it had been situated on a peninsula formed by a doubling of the river. We viewed the mounds or terraces, on which formerly stood their town house . . . and a little behind these . . . is a vast artificial terrace of four square mound, now seven or eight feet higher than the common surface of the ground; in front of one square or side of this mound adjoins a very extensive oblong square yard . . . sunk a little below the common surface, and surrounded with a bank or narrow terrace, formed with the earth thrown out of this yard at the time of its formation: The Creeks or present inhabitants have a tradition that this was the work of the ancients. . . .

Later, Bartram added this significant comment: "But in all the region of the Muscogulge country, South-West from the Oakmulge river quite to the Tallapoose . . . I saw . . . but one instance of the tetragon terraces, which was at the Apalachucla old town on the west banks of that river: here are yet remaining . . . vast four square terraces. Those Indians have a tradition that these remains are the ruins of an ancient Indian town and fortress." (Fretwell 1956: 63-65).

Bibliography

Anonymous

1940 Preliminary Report of Alabama Museum W. P. A. Archaeological Laboratory on Coosa Site. Filed at Mound State Monument.

Arnade, Charles W.

1959 *The Siege of St. Augustine in 1702,* pp. 9-11. University of Florida, Gainesville.

Barnwell, Joseph W.

1907 Dr. Henry Woodward, The First English Settler in South Carolina, and Some of His Descendants. *South Carolina Historical and Genealogical Magazine,* Vol. 8, January, 1907.

Battle, H. B.

1921 Sawokli Town. *Arrow Points,* Vol. 3, No. 5, pp. 81-82. Alabama Anthropological Society, Montgomery, Alabama.

Bolton, Herbert E.

1916 The Mission in the Spanish-American Colonies. *American Historical Review,* Vol. 23, October, 1916, p. 47.

1925 Spanish Resistance to the Carolina Leaders in Western Georgia (1680-1704). *Georgia Historical Quarterly,* Vol. 9, June, 1925, p. 122.

1925 *Arredondo's Historical Proof of Spain's Title to Georgia.* University of California, Berkeley.

Bolton, Herbert E. and Thomas Maitland Marshall

1920 *The Colonization of North America, 1492-1783.* New York.

Bolton, Herbert E. and Mary Ross

1925 *The Debatable Land,* pp. 34, 48, 54, University of California Press, Berkeley.

Boyd, Mark F., editor

1948 Don Pablo de Hito Salazar to the Queen, in "Enumeration of Florida Spanish Missions in 1675." *Florida Historical Quarterly,* Vol. 27, October, 1948, pp. 184-186.

Brannon, Peter A.

1909 Aboriginal Remains in the Middle Chattahoochee Valley of Alabama and Georgia. *American Anthropologist.* New Series, Vol. 9, No. 2, pp. 186-198.

1924*a* Russell County Place Names. *Arrow Points,* Vol. 8, No. 1, January, 1924, pp. 5-13, Alabama Anthropological Society, Montgomery.

1924*b* Cochcalechke Creek. *Arrow Points,* Vol. 9, No. 5, November, 1924, pp. 67-68, Alabama Anthropological Society, Montgomery.

1925 Casihta of the Lower Creeks. *Arrow Points,* Vol. 10, No. 3, March, 1925, pp. 44-45, Alabama Anthropological Society, Montgomery.

1930 Coweta of the Lower Creeks. *Arrow Points,* Vol. 17, No. 4, pp. 37-47, Alabama Anthropological Society, Montgomery.

Bullen, Ripley P.

1948 Personal Communication, December 31.

1950 An Archaeological Survey of the Chattahoochee River Valley in Florida. *Journal of the Washington Academy of Sciences,* Vol. 40, No. 4.

Caldwell, Joseph R.

1952 The Archeology of Eastern Georgia and South Carolina. *Archeology of Eastern United States,* edited by James B. Griffin, pp. 312-321. University of Chicago Press, Chicago.

1955*a* Cherokee Pottery from Northern Georgia. *American Antiquity,* Vol. 20, No. 3.

1955*b* Investigations at Rood's Landing, Stewart County, Georgia. *Early Georgia,* Vol. 2, No. 1, pp. 22-49. Society for the Preservation of Early Georgia History, Calhoun.

1958 Trend and Tradition in the Prehistory of the Eastern United States. *Illinois State Museum Scientific Papers,* Vol. 10. Memoirs of the American Anthropological Association, No. 88.

Caldwell, Joseph R., Catherine McCann, and Frederick Hulse

1941 *Irene Mound Site, Chatham County, Georgia.* University of Georgia Press, Athens.

Cambron, James W. and David C. Hulse

1960 The Transitional Paleo-Indian in North Alabama and South Tennessee. *Journal of Alabama Archaeology,* Vol. VI, No. 1, University, Alabama.

Caughey, John Walton

1938 *McGillvary of the Creeks.* University of Oklahoma Press, Norman.

Chase, David

Personal Communication.

Chatelaine, Verne Elmo

1939 The Defenses of Spanish Florida 1565-1763. *Carnegie Institution Publication* No. 511. Washington.

Coe, Joffre L.
1952 The Cultural Sequence of the North Carolina Piedmont. In *Archeology of Eastern United States,* edited by James B. Griffin. University of Chicago Press, Chicago.

Collections of the South Carolina Historical Society
1897 5 Vols. Richmond, Virginia.

Connor, Jeannette Thurber, trans.
1923 Pedro Menéndez de Avilés Adelantado Governor and Captain-General of Florida. Memorial by Gonzalo Solís de Merós. First Published in *La Florida, su Conquista y Colonización por Pedro Menéndez de Aviles,* by Eugenio Ruidìaz y Caravia. Florida Historical Society, Deland.

Crane, Verner Winslow
1928 *The Southern Frontier 1670-1732,* pp. 17, 36. Duke University Press, Durham.

DeJarnette, David L. and A. T. Hansen
1960 Archaeology of the Childersburg Site. *Florida State University, Notes in Anthropology, No. 6.*

DeJarnette, David L., Edward B. Kurjack, and James W. Cambron.
1962 The Stanfield-Worley Bluff Shelter Excavation. *Journal of Alabama Archaeology,* Vol. VIII, Nos. 1 and 2. University, Alabama.

Esquemeling, John
1910 The Buccaneers of America A True Account of the Most Remarkable Assaults Committed of Late Years Upon the Coasts of the West Indies By the Buccaneers of Jamaica and Tortuga (Both English and French) Wherein are Contained More Especially the Unparalleled Exploits of Sir Henry Morgan Who Sacked Porto Bello, Burnt Panama, etc. Swan Sonnenschein and Company, Ltd., London.

Fairbanks, Charles H.
1950 A Preliminary Segregation of Etowah, Savannah, and Lamar. *American Antiquity,* Vol. 16, No. 2, pp. 137-151. Salt Lake City.
1955 The Abercrombie Mound, Russell County, Alabama. *Early Georgia,* Vol. 2, No. 1, pp. 13-19, Calhoun.
1961 The Check Stamped Series. *Proceedings* of the Sixteenth Southeastern Archaeological Conference.
1961 Personal Communication.

Fenenga, Franklin and Barbara W. Fenenga
1945 An Archaeological Survey of the Vicinity of Columbus, Georgia. Unpublished manuscript.

Fretwell, Mark E., editor
1956 Two Early Letters from Alabama. *The Alabama Review,* Vol. 9, January, 1956, p. 59.

Goggin, John N.
1953 Seminole Archaeology in East Florida (Abstract). Newsletter, Southeast Archaeological Conference 3, No. 3, pp. 16, 19. Baton Rouge, Louisiana.

1957 Spanish Majolica in the New World, Types of the 16th to 18th Centuries in the Caribbean Region. Ms. to be published by the Middle American Research Institute, Tulane University, New Orleans.

1961 *The Spanish Olive Jar, An Introductory Study.* Yale University Publications in Anthropology, No. 57-64.

Great Britain, Public Record Office

1928-47 *Records in the British Public Record Office Relating to South Carolina 1663-1710,* p. 104. Foote and Davis Company, Atlanta.

Griffin, James B.

1943 The Analysis and Interpretation of the Ceramic Remains from Two Sites near Beaufort, South Carolina. *Anthropological Papers* No. 22, *Bureau of American Ethnology,* Bulletin 133.

Griffin, John W.

1950 Test Excavations at the Lake Jackson Site. *American Antiquity,* Vol. 16, No. 2, pp. 99-112. Salt Lake City.

Haag, William G., editor

1940*a* Newsletter. Southeastern Archaeological Conference. Vol. 2, No. 1. Mimeographed.

1940*b* Newsletter. Southeastern Archaeological Conference 2, No. 2, p. 9. Lexington, Kentucky. Mimeographed.

1948 Newsletter. Southeastern Archaeological Conference. Vol. 9, No. 2. Mimeographed.

Hamilton, T. M.

1960 Some Gun Parts From Eighteenth Century Osage Sites. Indian Trade Guns. *The Missouri Archaeologist,* Vol. 22, December, 1960. Columbia.

Harris, Walter A.

1958 *Here the Creeks Sat Down,* pp. 142-143. The J. W. Burke Company, Macon.

Hawkins, Benjamin

1848 A Sketch of the Creek Country in 1798 and 99. *Georgia Historical Collections,* Vol. 3, Part 1, pp. 19-85. Savannah.

1916 Letters of Benjamin Hawkins. *Georgia Historical Society Collections,* Vol. 9. Savannah.

Heimlich, Marion D.

1952 Guntersville Basin Pottery. *Geological Survey of Alabama, Museum Paper 32.*

Howard, Edgar B.

1935 Evidence of Early Man in North America. *The Museum Journal,* Vol. 24, Nos. 2-3. University of Pennsylvania Museum, Philadelphia.

Hudson, Paul

1960 Personal Communication to John Griffin, July 1960.

Huscher, Harold

1959 *Appraisal of the Archaeological Resources of the Walter F. George Reservoir Area, Chattahoochee River, Alabama and Georgia.* River Basin Surveys, Smithsonian Institution.

Jennings, Jesse D. and Charles Fairbanks
1939 Type Descriptions Swift Creek Complicated Stamp and Mossy Oak Simple Stamp. *Newsletter.* Southeastern Archaeological Conference, Vols. 1 and 2. Lexington, Kentucky.

Josselyn, Daniel W.
1952 Ground-base Projectile Points. *Tennessee Archaeologist,* Vol. 8, No. 3, pp. 65-72. Tennessee Archaeological Society, Knoxville.

Kelly, Arthur R.
1938 A Preliminary Report on Archaeological Explorations at Macon, Georgia. *Bureau of American Ethnology, Bulletin 119,* pp. 1-68, Washington.

Kelly, A. R., Edward V. McMichael and James H. Kellar
1962 *Archaeological Explorations at the Mandeville Site 9 Clal.* University of Georgia, Athens.

Lanning, John Tate
1935 *The Spanish Missions of Georgia.* University of North Carolina Press, Chapel Hill.

Lewis, Clifford M., and Albert J. Loomie
1953 *The Spanish Jesuits Mission in Virginia 1570-1577.* University of North Carolina Press, Chapel Hill.

Lister, Robert H.
1953 The Stemmed Indented Base Point, A Possible Horizon Marker. *American Antiquity,* Vol. 18, No. 3, pp. 265. Salt Lake City.

McMichael, E. V. and J. H. Kellar
1959 Archaeological Salvage in the Oliver Basin. University of Georgia, Athens.

Moore, C. B.
1907 Mounds of the Lower Chattahoochee and Lower Flint River. *Journal of Academy of Natural Sciences of Philadelphia,* 2nd Series, Vol. 13, Part 3, pp. 426-456. Philadelphia.

Morse, Dan and Phyllis Morse
1960 The Williams Site. *The Florida Anthropologist.* Vol. 13, No. 4, December 1960.

Pearson, Fred Lamar, Jr.
1963 Anglo-Spanish Rivalry in the Georgia Country 1670-1704. Ms., Master's Thesis, University of Alabama, pp. 19, 68.

Quiroga, Diego de
1688 *Quiroga to the King, April 15, 1688.* Archivo General de Indias 54-5-12. Seville.
1689 *Quiroga to the King, September 29, 1689.* Archivo General de Indias 54-5-12. Seville.
1690 *Quiroga to the King, June 8, 1690.* Archivo General de Indias 54-5-12. Seville.
1692 *Quiroga to the King, April 10, 1692.* Archivo General de Indias 54-5-13. Document No. 32. Seville.

Ross, Mary
1930 With Pardo and Boyano on the Fringes of the Georgia Land. *Georgia Historical Quarterly,* Vol. 15, December, 1930, p. 283.

Salley, Alexander Samuel, Jr.
1911 *Narratives of Early Carolina 1650-1708.* New York.

Sears, William H.
1951 Excavations at Kolomoki, Season 2-1950. *University of Georgia Series in Anthropology,* No. 3. Athens.

Sears, William H. and James B. Griffin
1950 Type Descriptions and Essays. In *Prehistoric Pottery of the Eastern United States.* Ann Arbor, Michigan.

Serrano y Sanz, Manuel, editor
1912*a* Carta Á. S. M. Del Gobernador de San Agustín De La Florida Don Diego de Quiroga San Agustín, 1° de Abril de 1688. In *Documentos Históricos De La Florida y La Luisiana. Siglos XVI al XVIII,* p. 220. Biblioteca de Los Americanistos, Madrid.
1912*b* Carta del Teniente de Apalache, Antonio Matheos al Señor Governador y Capitán-General De La Florida, acerca de Haber Vuelto Los Ingleses a La Provincia de Apalachicoli, San 19 de Mayo de 1606 [1686]. In *DocumentosHistóricos De La Florida y La Luisiana. Siglos XVI al XVIII,* p. 195. Biblioteca de Los Americanistos, Madrid.

Shea, John Gilmary
1886 *The Catholic Church in Colonial Days,* p. 163. The Mershon Company Press, Rahway.

Smith, Hale G.
1948 Two Historical Archaeological Periods in Florida. *American Antiquity,* Vol. 13, No. 4, Pt. 1, pp. 313-319. Menasha.

Swanton, John
1922 Early History of the Creek Indians and Their Neighbors. *Bureau of American Ethnology, Bulletin 73.* Washington.

Van Doren, Mark, editor
1928 *The Travels of William Bartram.* New York.

Walker, Anne K.
1941 *Backtracking in Barbour County.* Dietz Press, Richmond, Virginia.

Webb, William S. and David L. DeJarnette
1948*a* The Flint River Site, Ma°48. *Alabama Museum of Natural History Museum Paper 23.*
1948*b* Little Bear Creek Site, Ct°8. *Alabama Museum of Natural History Museum Paper 26.*

Willey, Gordon R.
1945 The Weeden Island Culture: A Preliminary Definition. *American Antiquity,* Vol. 10, No. 3, pp. 225-254. Menasha.
1949*a* Archaeology of the Florida Gulf Coast. *Smithsonian Miscellaneous Collections,* Vol. 113. Washington.
1949*b* The Florida Indian and His Neighbors: A Summary. In *The Florida Indian and His Neighbors,* pp. 139-167. Inter-American Center, Rollins College, Winter Park.

Willey, Gordon R. and Richard W. Woodbury
1942 A Chronological Outline for the Northwest Florida Coast. *American Antiquity,* Vol. 8, No. 3, pp. 232-254. Menasha.

Witthoft, John
1954 A Note on Fluted Point Relationships. *American Antiquity,* Vol. 19, No. 3, pp. 271-273. Salt Lake City.

Wright, Irene A., trans.
1920 Spanish Policy toward Virginia 1606-1612; Jamestown, Ecija, and John Clark of the Mayflower. *American Historical Review,* Vol. 25, pp. 449-478.

Index